Luiz Gustavo Batista Ferreira

Advances in Yields and Zoning Risks of Soybean crop

Luiz Gustavo Batista Ferreira

Advances in Yields and Zoning Risks of Soybean crop

Noor Publishing

Cover image: www.ingimage.com

Publisher:
Noor Publishing
is a trademark of
Dodo Books Indian Ocean Ltd. and OmniScriptum S.R.L publishing group

120 High Road, East Finchley, London, N2 9ED, United Kingdom
Str. Armeneasca 28/1, office 1, Chisinau MD-2012, Republic of Moldova, Europe
Printed at: see last page
ISBN: 978-620-7-47917-7

Table of Contents

Author:

Luiz Gustavo Batista Ferreira born in Londrina, Brazil. Agronomist Engineer graduated at State University of Londrina, with a very good grade; Master of Science in Conservation Agriculture; Autor of several articles published around the world; Ph.D student in Agronomy at State University of Londrina.

INTRODUCTION

The soybean crop is the most important crop in Brazil being considered responsible for around 30 % of global production in 2021 (SAMPAIO et al., 2021), covering an area of more than 39 M hectares in the same crop season (EMBRAPA, 2021; SANTOS et al., 2021).

According to USDA (2023), for the crop season 2022/2023, soybean crop occupied an area with almost 44 M hectares and range a yield of 3.6 $T.ha^{-1}$. The main area of productions of soybean in Brazil are South and Midwestern, areas with higher latitudes, >15°S, and thanks the genetic breeding and advances in agricultural managements, the North and Northeast have important precipitation for soybean productions in Brazil, even though are considered areas with low latitude, <15°S (SAMPAIO et al., 2021). Soybean is cultivated between September to March (Spring-Summer crop season) (FERREIRA et al.,2020).

However, although the recent scientific and technological advances and performance of soybean in soils with low fertility thanks to recent advances in cultivars, the climate has the most important element for variability for crops yields, including soybean yields, due 80 % of yield variability come from the climate (CARAMORI et al., 2008; CALDANA et al., 2019; FERREIRA et al., 2020; CALDANA et al., 2021).

In order to breed yield performances and to reduce climatic risks, the applicability of studies on agrometeorological elements has been applied on a global scale, helping the farmer in decision-making, in addition, studies can reduce the soybean yields gaps (GODFRAY et al., 2010; SENTELHAS et al., 2015; MARTINS et al., 2017; CALDANA et al., 2021; FERREIRA et al., 2020). For this context, Battisti et al. (2017) suggest to improve the agricultural management strategies to improve crop resilience to climate factors, even in areas with considered as higher risks for cultivation (HEINEMANN et al., 2016; ZANON, STRECK, GRASSINI, 2016).

Appropriated crops managements are recommended to increase the soybean yields, even in areas with higher climate risks, thus, improving soybean crop resilience and there is no necessity to deforestation (AGOVINO et al., 2019; SAMPAIO et al., 2021). Thanks to crop managements, soybean can use introduced in new environments, and, in this context, dates of sowing and maturity group are very important (SAMPAIO et al., 2021)

For Sampaio et al. (2021), soybean yield are results of the interactions of atmospheric, soil and crop managements, thus, considering: air temperature, rainfall, photoperiod, soil water availability, plant density, sowing date and maturity group. Ferreira et al. (2020) suggest the sowing date of soybean can impact the yields, due the risks of droughts periods, mainly in November, in Paraná state, South of Brazil. Meotti et al. (2012) pointed out the sowing dates are important interfere in water relations and the variables mentioned, that can interfere in cultivars yields.

The yield gap of soybean can be avoided, by defining the best plant density for each combination of sowing date and maturity group for environment (BATTISTI, BENDER, SENTELHAS, 2018). For this context, Sampaio et al. (2021) recommended crop models to help the farmer in decision-making, considering climate, maturity groups, sowing dates and plant density

Agrometeorological crop models can help the farmer for decision-making and for agricultural planning, such as crop adaptation, chose the best cultivar for a given local, best time to sowing, crop monitoring, plant density, forecasting and even for control the pest and disease incidences. Tools, such as crop models, has been developing aiming to explain the great interactions between crops, crops management and weather conditions (BATTISTI, BENDER, SENTELHAS, 2018; BATTISTI, 2016).

Rainfall are the most important element and the most significant meteorological attribute in tropical and subtropical areas, where its variable distribution and occurrences of extreme periods between droughts and rainfall in excess (CALDANA et al., 2020; FERREIRA et al., 2020). Study of climate phenomena can contribute to improvements in socioeconomic decision-making, minimizing impacts (CALDANA et al., 2022).

According to Ferreira (2017), among the climatic phenomena responsible for rainfall variability in southern Brazil is ENSO (El Niño Southern Oscillation). ENSO is characterized by the heating or cooling of the Equatorial Pacific waters, that is, anomalies in the SST - Sea Surface Temperature and has a hot phase and a cold phase and, in addition, PDO (Pacific Decadal Oscillation). The warm phase is called El Niño and the cold phase, La Niña (CUNHA et al., 2001;

FERREIRA, 2017) and this phenomenon are more intense due the climate change, thus, are important studies which consider these phenomenon, the climate change and how these can interfere in soybean yield and help the farmer to choose the best data of sowing.

Soybean crop models have a great importance to help the decision-making of the farmer, in addition, there studies using crop models for soybean crop aiming reducing yields gaps and for plants density (BATTISTI, BENDER, SENTELHAS, 201; SENTELHAS et al., 2015; TEIXEIRA et al., 2018; SAMPAIO et al., 2021).

In this context, this study intends to use the crop simulation modeling (CSM)-CROPGRO-Soybean is able to simulate development and growth due the soybean data sowing, to define the best strategies considering the sowing dates and the context of climate changing scenarios published by IPCC, new crop managements can be adopted to minimize the climatic risks.

REVIEW

SOYBEAN: CULTIVATION AND AGRICULTURAL MANAGEMENT

Soybean is a plant native from China, belonging to the Fabaceae family, subfamily Faboideae, genus Glycine and species *Glycine max* (L.) Merrill, being considered one of the oldest crops in the world. Soybean is main crop cultivated in Brazil (FERREIRA et al., 2020).

In Brazil, the soybean crop was implemented in 1882. In 1914, it was introduced in the Rio Grande do Sul State, due to favourable climatic conditions, especially the photoperiod. The implementation of soybean genetic breeding programs made it possible to advance the crop to regions with low latitudes (FERREIRA, 2017).

Soybean is one of the most cultivated and ancient species in the world. It has conducted with a high technological level in all its operations, constituting a "commodity" of great commercial importance for the development of Brazilian agriculture, being grown in all regions. Its great economic importance in the scenario Brazilian is due to its high volume produced.

Soybean phenology is admitted as vegetative (V) and reproductive (R), this botanical parameter is so important to understanding the interactions between plant and environment, to help the farmer in decision-making, for example, choosing the best moments to sowing and even for crops simulations models (FERREIRA et al., 2020; SAMPAIO et al., 2021). Ferreira et al. (2020) describes the cycle of soybean from VE (emergency) to R8 (maturation, harvest).

Figure 1. Cycle of soybean, where V are vegetative stages and R are the reproductive ones

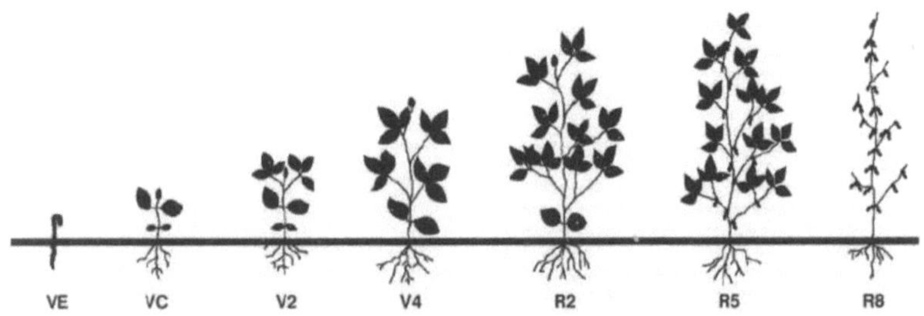

Source: FERREIRA, L.G.B.; CALDANA, N.F.S.; MARTELÓCIO, A.C.; COSTA, A.B.F.; NITSCHE, P.R.; CARAMORI, P.H. Rainfall variability and analysis of droughts periods risks during the soybean crop (*Glycine max* L.) in the Western of Paraná State, Brazil. **Brazilian Journal of Climatology** v.27, p.590-611, 2020.

For the USDA (2023) the productions of soybean, in the crop season 2021/2022 was 156.000 T, covering an area with 43.900 M ha and yield of 3.6 T.ha. Figure 2 shows the Brazilian crops productions.

Figure 2. Soybean Productions in Brazil.

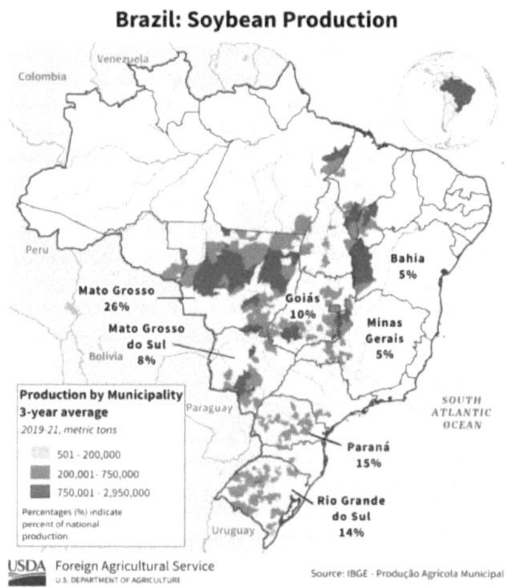

Source: USDA – UNITED STATES DEPARTAMENT OF AGRICULTURE: International Production Assessment Division. Brazilian soybean productions, 2023
https://ipad.fas.usda.gov/countrysummary/Default.aspx?id=BR&crop=Soybean

According to the Figure 2, soybean crop is cultivated in all regions of Brazil, from Rio Grande do Sul to Roraima, at the latitude Equator Line. The main regions of soybean cultivation is the Midwestern of Brazil (Mato Grosso, Goiás, Mato Grosso do Sul states and South of Brazil, especially Rio Grande do Sul and Paraná states (SAMAPAIO et al., 2021).

The cultivation of soybean in latitudes from the Midwestern to North and Northeast was possible due genetic breeding to developing new cultivars (FREITAS, 2011), in addition, crop management such as fertilization and correction of acids soils allow the soybean cultivation in North/Northeast, with participation almost 20 % of National production (SAMPAIO et al., 2021).

Another major factor that contributed to the expansion of soybeans in Brazil was the implementation of integrated pest, such fungicides, management, controlling the main insects economic damage the soybean crop (FREITAS, 2011). The Figure 3 exhibits the participation of each region in Brazil.

Figure 3. Soybean productions participation of each region in Brazil.

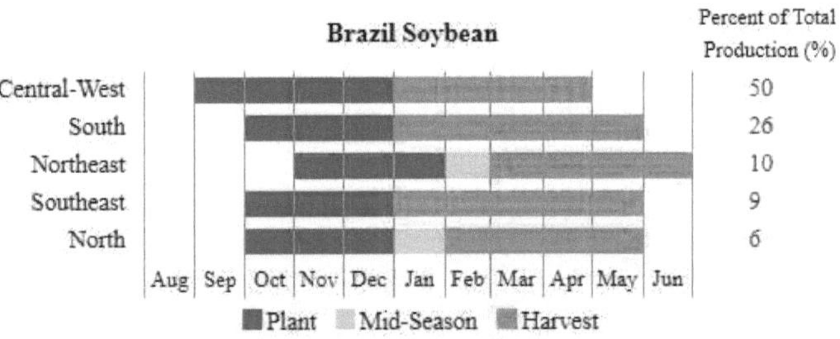

Source: USDA – UNITED STATES DEPARTAMENT OF AGRICULTURE: International Production Assessment Division. Brazilian soybean productions, 2023
https://ipad.fas.usda.gov/countrysummary/Default.aspx?id=BR&crop=Soybean

Soybean cultivation in Brazil has great social and economic value, due to the importance of its products, mainly bran, vegetable oil and its derivatives, destined for both the domestic and foreign markets, which results in significant job creation in the various sectors of the national economy (SENTELHAS et al., 2015; FERREIRA, 2017). For Battisti (2016) the soybean grain can be used to can be used to produce several types of products, such as refined oil for cooking, biodiesel, defatted flour, meal to animal feeding, isolated protein, and, also, as fresh food.

The mean actual yield (AcY) of soybean in Brazil is of 3,206 kg.ha^{-1} (SANTOS et al., 2021), and this value represents about half of yield observed in the best farm fields in Brazil (BATTISTI et al., 2018). The authors verified that soybean yield at farm level reached more than 5,000 kg.ha^{-1}, when evaluated 200 farm fields across Brazil. These levels are similar as those the Global Yield Gap Atlas reported as a potential yield for soybean in Brazil (SANTOS el al., 2021).

In general, soybean is cultivated during the Spring-Summer of South Hemisphere, between September and March in Brazil, Figure 4 (FERREIRA et al., 2020). Studies which contribute for decision-making and establish an appropriated agricultural planning and management, for example, the best moment for sowing, necessities of irrigation and cultivar choice (PATHMESWARAN et al. 2018; WIRÉHN, 2018; SANTI et al., 2018; DE SOUSA; DE SOUZA et al., 2018; TAYT'SOHN et al., 2018; AGOVINO et al., 2019).

Figure 4. Soybean sowing and harvest in Brazil.

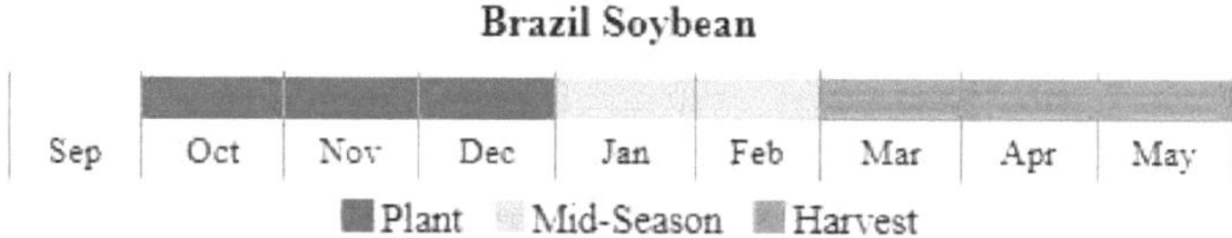

Source: USDA – UNITED STATES DEPARTAMENT OF AGRICULTURE: International Production Assessment Division. Brazilian soybean productions, 2023
https://ipad.fas.usda.gov/countrysummary/Default.aspx?id=BR&crop=Soybean

Soybean crop, as other crops in rainfed areas in Brazil, are higher dependent on the climate conditions, according to the Ferreira et al. (2020), even though the recent scientific and technology advances, the variability of meteorological conditions is still the most important factor for crops yields in rainfed areas, especially soybean crop during the reproductive stages, where the crop requires a necessity quantity of water for the plant physiology.

For Sentelhas et al. (2015) the main environmental factor responsible for yield losses in rainfed soybean is drought, which is responsible for approximately 74 % of yield gap, reducing yields, in average, by 30 %.

These yields variabilities exposed by Ferreira et al. (2020) and Sentelhas et al. (2015) in rainfed areas are explained by the phenomenon ENOS – El Nino South Oscillation (ENOS), with affects the rainfall variability (FERREIRA, 2017).

According to Ferreira (2017), among the climatic phenomena responsible for rainfall anomalies in South of Brazil is ENSO (El Niño Southern Oscillation). ENSO is characterized by the heating or cooling of the Equatorial Pacific waters, that is, anomalies in the SST - Sea Surface Temperature and has a hot phase and a cold phase, being the warm phase is called El Niño and the cold phase, La Niña (CUNHA et al., 2001).

Anomalies of extreme rainfall events are caused by an atmospheric circulation called Hadley, influencing the jet stream at an altitude of 10,000 m, in El Niño years, this jet blocks the atmosphere, causing cold fronts to be parked over southern Brazil, causing excess rainfall in southern Brazil. In La Niña years, the Hadley-type cell is weakened (CUNHA et al., 2001).

Normally, a reduction in rainfall is observed during La Niña phase, whereas an increase occurs during El Niño phase (FARIAS et al., 2007). Such climate variability has a possibility to become more frequent under climate change scenarios regions (LI et al., 2013; BATTISTI, 2016). Verifying the Figure 5 it is possible to compare the effect of ENOS on the soybean yields in Brazil.

Figure 5. Soybean yield in the crop seasons of 2010/11 (neutral year) (a) and 2011/12 (La Niña year) (b), in Brazil.

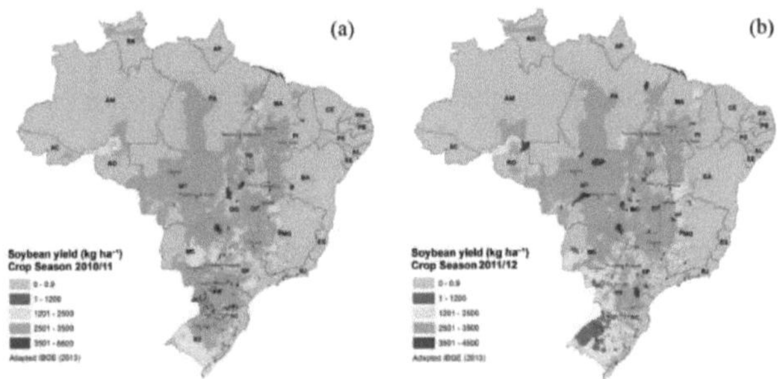

Source: BATTISTI, R. **Calibration, uncertainties and use of soybean crop simulation models for evaluating strategies to mitigate the effects of climate change in Southern Brazil**. Doctorate Thesis in Agronomy (Agricultural Systems Engineering). University of São Paulo "Luiz de Queiroz" College of Agriculture. 189 f., 2016. Available in:https://www.teses.usp.br/teses/disponiveis/11/11152/tde-03102016-162340/en.php

The soybean yield is result of the interactions between climate conditions, especially rainfall in rainfed areas and crop management, such as maturation group of cultivars (MG), plant density and sowing dates (FERREIRA et al., 2020; SAMPAIO et al., 2021).

Ferreira et al. (2020) and Ferreira (2017) observed the phenological reproductive stages were significant important for soybean yield at R2 (full bloom) or R5 (beginning seed). For Cunha et al. (2001), the two critical stages of soybean cultivation, in relation to water availability, are from the germination emergency and post-flowering. It is should be noted that these are considered important parameters for crop simulation models of soybean, considering the phenology, water availability and crop management (cultivars, plant density and sowing dates).

From this perspective, Steeke and Grabal (1997) exhibited soybean yield is directly connected with sowing dates, due the authors verified shortest cultivar showed less yield. The crop management parameter sowing dates is considered as very important for soybean crop phenology, and for yield, as consequence (RAHMAN et al., 2005; ZHANG et al., 2010; HU; WITRACK, 2012).

Helping the farmer in his decision-making to choose the better cultivar and, mainly, the best moment to sowing can improve soybean growth and development, and enhance the yield potential (HU; WITRACK, 2012; MEOTTI et al., 2012).

As mentioned, the climate is responsible for higher percentual, 80 %, of soybean yield in rainfed areas, then the other 20 % are about the crop management, especially plant density, sowing dates and cultivars, which considerer the MG due these factors interfere directly in biomass and the time of flowing (EGLI; CORNELIUS, 2009; HU; WITRACK, 2012; SENTELHAS et al., 2015).

These parameters are necessary to help the farmer in decision making to make the soybean resilience, in areas with low latitudes for example, or to decide the best moments to sowing for avoid risks of droughts periods, in case of Paraná state, reaching 15 % in November, with the objective to reduce the yields gaps due water availability and due crop management, thus, sowing the best moments can avoid this period during the soybean flowing, exhibiting less risks of yield gap due the available water (SENTELHAS et al., 2015; FERREIRA et al., 2020; SAMPAIO et al., 2021).

Parameters of soybean crop management, plant density, sowing dates and cultivars are important to increasing yield, instead of deforestation for example or even to explore areas of Northeast and part of North in Brazil, areas with limited the potential yield of soybean due the soil and especially climate conditions, due these areas are near the Equator line (SAMPAIO et al., 2021; TEIXEIRA et al., 2019).

The management needs to be adapted for different environmental conditions to reduce risks associated with climate and production costs (BATTISTI et al., 2020). Thus, crop management practices that can be used in new environments to improve yield and crop resilience, include sowing dates (HU; WIATRAK, 2012; SPEHAr, FRANCISCO, PEREIRA, 2015) and maturity group (BATTISTI et al., 2018; TEIXEIRA et al., 2019).

For Sampaio et al. (2021) the plant density of 40 plants.m^2 promote higher yield of soybean crop in crop models simulations, while a higher plant density can promote yields gaps due the higher quantity of diseases what reduces the phytosanitary control, this parameter is difficult to simulate and, generally, are not admitted in crop models simulations (SAMPAIO et al., 2021). Thus, the planning of maturity groups and sowing dates by sites showed to be important factors by the different response patterns, which can help to improve soybean yield in Brazil (SAMPAIO et al., 2021).

Photoperiod and temperature exert influence for soybean crop phenology, with reflections on plant height, and potential yield of crop productivity (JIANG et al., 2011). Late sowing can lead to loss of order of 30 to 50 % in grain yield, while sowings in the off-season can yield gap of 70 %, in relation to the recommended season, thus, the adoption of times of sowing that provide conditions to those required by plants is extremely important for soybean yields (MEOTTI et al., 2012).

Sowing date interfere in solar radiation interception during the critical periods for grain set were significantly and directly correlated with soybean yield (ANDRADE et al., 2002).

The influence of radiation was found to be reduced in late planted MG (maturations group) III and IV soybean cultivars (EGLI; BRUENING, 2000). Final biomass responded linearly to cumulative intercepted photosynthetically active radiation until it reached 400 MJ m^{-2} and it began to decrease with radiation levels exceeding 400 MJ m^{-2} (PURCELL et al., 2002). Asanome and Ikeda (2000) suggest that beyond the radiation, the sowing times affect de Nitrogen concentration in leaf of soybean.

Seed number per area was significantly correlated with crop growth rate during flowering stages (R1–R3) and pod set (R3–R5) stages (Seed and pod numbers had the greatest impact on grain yield, it is should be noted that these trends are more significant for indeterminate than determinate soybean (HU; WIATRAK, 2012). Weaver et al. (1991) found that these stages of soybean crop was reduced in later sowing dates for both indeterminate and determinate soybeans.

Heatherly (2005) reported that late sowing date reduced duration of both vegetative and reproductive growth stages of maturation group (MG) IV through VI soybean. He also indicated that the major difference among cultivars of different MGs was the length of vegetative (sowing to R1) rather than reproductive stages (R1–R8). The longer cycle (MG 8.8) was a better strategy for low latitude, where water deficit is lower and longer cycle results in higher potential yield (BATTISTI; SENTELHAS, 2019).

The maturity group and sowing dates are affected by interaction between local weather and plant density over yield due to soybean cycle duration and the time to achieve the optimum LAI (leaf area index) for maximum SR (solar radiation) interception (ZDZIARSKI et al., 2018).

The maturity groups and sowing date patterns by location occurred due to cycle duration and potential yield associated with climate (BATTISTI; SENTELHAS, 2019). The cycle duration defines the capacity of the crop to intercept solar radiation and, consequently, the potential yield (VAN ROEKEL et al., 2015).

Sowing at the best moment can avoid climatic risks, droughts periods in Brazil in rainfed areas or frosts risks in U.S.A, thus the crop management sowing dates is very important for soybean yields and minimize risks of yields gaps (HU; WIATRAK, 2012). In addition, Hu and Witrak (2012) suggested sowing time of soybean can interfere in other aspects such as soil temperatures, especially in cold places and the management of insects; the authors made a correlation between frustrations productivity and wrong dates to sowing.

Sowing dates for Brazilian conditions vary according to the regions and cultivars, presenting a range recognized mendable from October to December. It was observed, however, which the month of November has provided, in general, the best results in the of those who cultivate this crop more traditionally (NAKAGAWA, ROSOLEM; MACHADO, 1983).

Nakagawa, Rosolem and Machado (1983) have verified that the sowing time can affect different production, such as: number of pod.m^2 (CARTER, 1974) or per plant. In addition, there are other agronomic parameters, as number of grains per pod, number of seeds.m^{-2} and seed weight. It is should be noted that these agronomic parameters can be used in DSSAT simulations (TSUJI; HOOGENBOON; THORTON, 1998).

Weaver et al. (1991) indicated that indeterminate cultivars had less yield loss due to delayed sowing, although determinate soybean yielded higher than indeterminate plants. Plant height could be affected by sowing date and location (PEDERSEN; LAUER, 2003). Radiation use efficiency (RUE, intercepted photosynthetically active radiation in g dry matter MJ^{-1}) is an important function of crop productivity (MONTEITH, 1972).

CROPS PRODUCTIONS: YIELD TYPES FOR SOYBEAN CROP

The soybean yield is result of great interactions between crop management (sowing dates, plant density and maturity group of cultivars) and climate conditions (SAMPAIO et al., 2021). The crop management can be input in crop models simulations of soybean to study the interactions crop-climate-management, aiding reducing yields gaps (SENTELHAS et al., 2015; SAMPAIO et al., 2015).

For Sentelhas et al. (2015) soybean yield depends on determining, limiting and reducing factors, known as determining factor: solar radiation (SR), species, temperature (T), cultivar, plant population, plant density and sowing dates (SAMPAIO et al., 2021); irrigation (limiting and reducing factors); water deficit and SWHC - soil water hold capacity (limiting factors) and crop management as control of pests/diseases and soil management (reducing factors) (FISCHER, 2015; SENTELHAS et al., 2015; SENTELHAS et al., 2016; LIU et al., 2016).

To comprehend precisely the soybean yields and its climatological interactions, it is necessary to understand the three different types of yield: the potential yields (Yp) and attainable yields (Ya), and average real yield (Yr) (FERMONT et al., 2009; BINDRABAN et al., 2000; VAN ITTERSUM et al. 2013; SENTELHAS et al. 2015; NENDEL et al. 2014; BATTISTI et al. 2017b).

Figure 6. Conceptions of determined factors of production

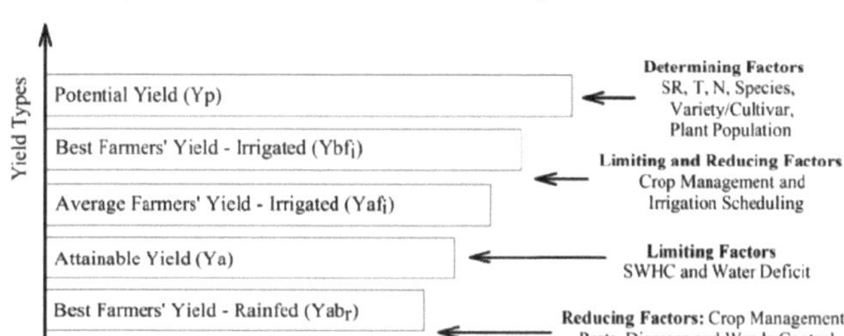

Source: SENTELHAS, P.; BATTISTI, R.; CÂMARA, G. M. S.; FARIAS, J. R. B.; HAMPF, A. C.; NENDEL, C. The soybean yield gap in Brazil–magnitude, causes and possible solutions for sustainable production. **The Journal of Agricultural Science**, p. 1-18, 2015.

Depending on the local, each researcher can adapt or complement the yield definitions mentioned for a given local or regional experiment, or if the experiment is rainfed or irrigated, as Sentelhas et al. (2015) exhibited in the Figure 6. The authors exhibited complementary definitions for crop modelling: Best farmers yield irrigated (Ybf_i); average farmers yield irrigated (Yaf_i); attainable yield (Ya); best farmers yield – rainfed (Yab_r) and average farmers yield rainfed (Yaa_r)

Lobell, Cassman and Field (2009) established concepts for their specific study: Modelled yield potential; experiment yield; maximum farmer yield and average farmer yield, thus, the researchers suggested that this explain the yield gap (YG_M), experiment-based yield gap (YG_E), and farmer-based yield gap (YG_F) (LOBELL; CASSMAN; FIELD, 2009), as can be seen in the Figure 7.

Figure 7 - A conceptual description of average farmers yields and three measures of yield potential

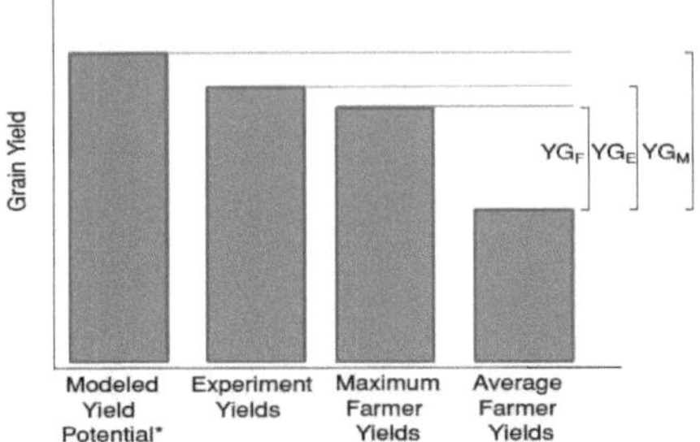

Source: LOBELL, D. B., CASSMAN, K. G.; FIELD, C. B. Crop yield gaps: their importance, magnitudes, and causes. **Annual Review of Environment and Resources**, v. 34, p. 179-204, 2009.

The authors Van Ittersum et al. (2013) proposed different levels of productions and their respective limited factors, as potential yield (Yp) limited by CO_2, radiation, temperature and cultivar features; water-limited (Yw) limited by water and nutrients and actual yield, limited by crops pests/diseases.

Figure 8 – Productions situations and yields levels (a) and the difference between the average yields and 80 % of potential yield (b)

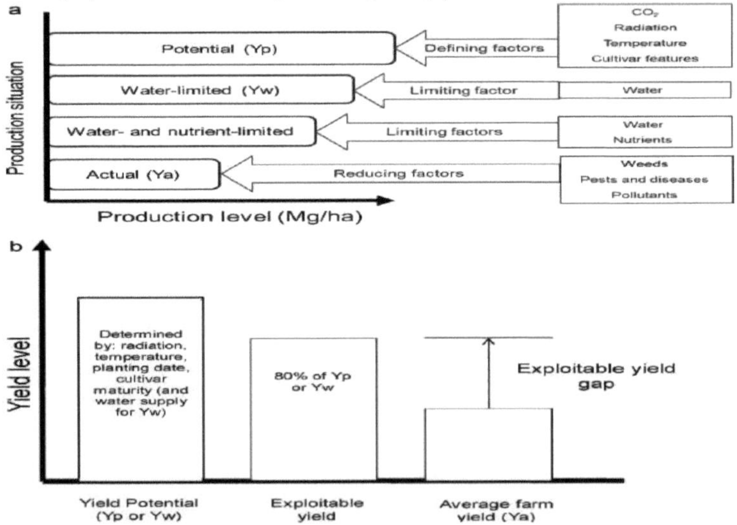

Source: VAN ITTERSUM, MARTIN K.; CASSMANB, K. G.; GRASSINI, P. B.; JOOST W. A.; TITTONELLI, P. C.; HOCHMAND, Z. Yield gap analysis with local to global relevance - a review. **Field Crops Research**, v. 143, p. 4-17, 2013. Access in: 10 Sep, 2021

According to Van Ittersum et al. (2013), Yp is determined solely by solar radiation, temperature, atmospheric CO_2, interception of global solar radiation by the crop canopy and genetic characteristics that govern the length of the species' growth period in that location.

The potential yield (Yp) is defined as the maximum yield that a crop can have under a certain climatic condition, without water and nutritional limitations (limiting factors) and occurrence of pests, diseases and weeds (reducing factors), this level of productivity is theoretical, being a function of genetic material, plant population, solar radiation, photoperiod and air temperature, which are considered as determining factors (PONTI; RIJK; VAN ITTERSUM, 2012).

Fischer (2015) suggested the potential yield (Yp) is the maximum biological yield expected with the best variety adapted to the agricultural environment, the best agricultural management of agronomic parameters inputs and the absence of manageable biotic and abiotic stresses. For Evans (1993) and Van Ittersum et al. (2013) is the yield of a crop cultivar when grown with water and nutrients non-limiting and biotic stress effectively controlled.

The potential yield (Yp) can be estimated or determined through simulation models, where it is assumed that there is perfect agronomic management, without the presence of limiting factors; or through field experiments, it is necessary to ensure complete control of the environment so that Yp is not underestimated (LOBELL; CASSMAN; FIELD, 2009).

The potential yield (Yp), in theory, not depends on the soil properties, assuming the required water and nutrients can be added through by the agricultural management, thus, in areas without major soil constraints, the potential yield is the most relevant benchmark for irrigated systems or systems in humid climates with adequate water supply to avoid water deficits, this parameter (Yp) can be defined as water-limited yield (Yw), however, it is should be noted that crop growth is also limited by water supply, and hence influenced by soil type (water holding capacity and rooting depth) and field topography (VAN ITTERSUM et al., 2013).

The potential yield and water-limited yield are calculated considering the recommended sowing dates, planting density and cultivar (which determines growing period to maturity). In addition, sowing dates and cultivar maturity are specified to fit the cropping system, because the cropping system “context” is critically important in for growth duration, particularly in tropical and semi-tropical environments where two or even three crops are produced each year on the same area (VAN ITTERSUM et al., 2013).

The attainable yield (Ya) represents the yield achievable using current best-known technology and management techniques at a given time and in a given ecosystem. This type of yield can be estimated using crop simulation models, which use climate and crop data to estimate growth, development and yield (LOBELL; CASSMAN; FIELD, 2009; BHATIA et al., 2008; SENTELHAS et al., 2015; LIU et al., 2016). The attainable yield (Ya) represents the difference between average yields and 80 – 85 % of potential yield (Yp) (LIU et al., 2016).

As meteorological data vary during the year, potential yield (Yp) and attainable yield (Ya) also vary, according to the date of implementation of the crop, which makes these yields also dependent on the location and date of planting (VAN ITTERSUM et al., 2013).

In addition, the population of chosen plants also directly affects their productive characteristics, due to its interactions to accumulate dry matter (DUARTE, 2018). This accumulation of mass and productivity will be maximum when it is possible for the population to cover the ground in order to intercept the maximum amount of solar radiation in a shorter period of time (LOBELL; CASSMAN; FIELD, 2009).

Bindraban et al. (2000) suggested the actual farmer yield (Yr) is obtained as a function of the efficiency of nutritional management, pests, diseases, weeds and soil preparation (production-reducing factors), and is also a function of the biophysical and socioeconomic conditions of the growing region.

For this context, the technological level employed by farmers becomes decisive for defining the level of soybean yield to be obtained. Thus, for the crop to have maximum yield in the production environment in which it is inserted, it is essential that the factors that limit and reduce its production (water, nutrients, pest management, diseases and weeds) are not neglected (SENTELHAS et al., 2016).

Pets and diseases must be considered in crop management (TAKATSU; FUKUDA, 1990 FIALHO et al., 2001 SCHIMITT, 2002 ALBUQUERQUE et al., 2008). This parameter (pests and plant diseases) were not considered in crop simulations models made by Sampaio et al.(2021). According to the authors, their result of crop simulation model (CSM) for soybean crop do not considered the incidences of pests and diseases, but they point out the higher density of plant can make incidences of phytosanitary issues for soybean crop.

The yield gap (Yg), conception with key importance to crop modelling, is the is the difference between average and is are estimated by the difference between Yp (irrigated crops), or Yw (rainfed crops) and actual average farmer yields (Yr) (LOBELL; CASSMAN; FIELD; SENTELHAS et al., 2015) or simply the difference between the potential yield (Yp) and average real yield (Yr) obtained by farmer (SENTELHAS et al., 2015).

The first step to mitigate yield gaps for soybean crop is to identify breeding cultivars with resistance or non-resistance for water deficit (LEHMANN et al., 2013). Cultivars characteristics, such as deeper and denser root system (BORTOLUZZI et al., 2014), can be useful in the context of soil water conservation (SINCLAIR et al., 2007; SINCLAIR et al. 2008; GILBERT et al., 2011), thus, a breed cultivar are higher relevant to increase the crops productions.

To understand the variability on the crops productions due by water deficit or crop management, it is necessary to understand the concept of yield gap (Yg) (SENTELHAS et al., 2015). The yield gap is the difference between different yield types for a specific condition considering spatial and temporal scales. Sentelhas et al. (2015) pointed out some important considerations that there are different types of yield gaps, for example, the total yield gap and yield gap due the crop management (SENTELHAS et al., 2016).

The total yield gap (YG_T), what consider all determining, limiting and reducing mentioned by Sentelhas et al. (2015), is the difference between the potential yield (Yp) and average real farmer yield (Yr). The yield gap (YG_{WD}) that consider only the water deficit is the difference between potential yield (Yp) and attainable yield (Ya) and at last, the yield cap (YG_M) caused by crop management is the difference between attainable yield (Ya) and average real farmer yield (Yr) (SENTELHAS et al., 2015).

YGM, as mentioned, refers to yield gaps caused by crop management. For this, Sampaio et al. (2021) pointed out the main agricultural practices for crop management are plant density, cultivars (considering maturation group of soybean) and sowing dates of soybean crop (TEIXEIRA et al., 2019; SAMPAIO et al., 2021).

The yield gap (Yg) may be due to water deficit (YG_{WD} = Yp - Ya) and attributed to the agricultural management adopted in the crop (YG_M = Ya - Yr) (FISCHER, 2015; LIU et al., 2016).

These concepts with yields types are important to reducing the yields gaps of soybean crop, giving important strategies to increase soybean yield, using less area as possible, in context of climate change and food security, or creating soybean resilience for new environments (LOBELL; CASSMAN; FIELD, 2009; SENTELHAS et al., 2015; SAMPAIO et al., 2021).

CROP MODELLING SIMULATIONS

In agriculture, the great advantages and importance of crop simulation models, are the possibility to evaluate experiments in macro scale, as this Thesis purposes and Battisti (2016) and Battisti, Bender and Sentelhas (2019) made in their studies. Once determined the areas of the experiments, are necessary calibrate and validate the model (JONES et al., 2003; EGLI; CORNELIUS, 2009; HUNT; BOOTE, 1998).

Crop simulation models are tools to estimate crops development, growth and productivity and, in addition, aims to evaluate the influence of the meteorological conditions on crop production, showing confinable results, considered as a remarkable tool to help the farmer in decision-making and reducing yield gaps (REYNOLDS, 1979; MONTEIRO; SENTELHAS, 2014; MARTINS, 2012; BATTISTI, 2016; BATTISTI, BRENDER, SENTELHAS, 2019).

Comprehending the crop modelling simulations is considered a key strategy associated with adopting the best crops management to make crop resilience, mainly in areas with higher climatic risks (BATTISTI et al., 2018; BATTISTI et al., 2017a; DO RIO et al., 2015; HEINEMANN et al., 2016; ZANON et al., 2016).

The development of applied crop models already started at least 65 years ago (JOYCE; KICKERT, 1987). Peart and Barret (1976) exhibited a pioneer study in modelling of crops; some years after, the authors Sakamoto and LeDuc (1981) showed a panorama of the historical development of crop meteorological parameter models. Getz and Gutierrez (1982) describe a few historical papers on applied modelling the plant physiology and its impact on the ecological theory, biology of populations and on the development of new techniques of resource management and agroecosystem modelling.

Some years later, the program DSSAT, Decision Support System for Agrotechnology Transfer, was first released in 1998 and the program is used to estimate production, resource use, and risks associated with different crop production practices and with a key purpose to aid the farmer in complex decision-making in field (TSUJI; HOOGENBOON; THORTON, 1998).

The program DSSAT, is used to simulate crops productions to verify, for example, the best time to sowing (BATTISTI, BENDER; SENTELHAS, 2019). For this context, DSSAT can applied, for example, to soybean crop (SENTELHAS et al., 2015).

Battisti, Brender, Sentehas (2019) used the System for Agrotechnology Transfer (DSSAT) for soybean modelling in Brazil, according to the authors, the use of DSSAT can be transforming information in useful knowledge to apply in agriculture, in addition, showing significant contribution to scientific and technology advances (BATTISTI, BENDER; SENTELHAS, 2019).

For simulation of this study were used Decision Support System for Agrotechnology Transfer – DSSAT, to verify the best time to sowing soybean and comparing with the farmer's National champion of soybean producer.

DSSAT is used to estimate production, resource use, and risks associated with different crop production practices and with a key purpose to aid the farmer in complex decision-making in field, showing crop-soil simulation models, data bases for weather, soil, and crops, and strategy evaluation programs and provides a framework for scientific cooperation through research to enhance its capabilities and apply it to research questions (TSUJI; HOOGENBOON; THORTON, 1998).

The utility of this system depends on the ability of the crop models to provide realistic estimates of crop performance for a wide range of environment and management conditions and on the availability of data required to operate the models and can be applied independent of location, season and agricultural practices and researchers around the world can use DSSAT for simulation and help farmers to decision-making (TSUJI; HOOGENBOON; THORTON, 1998).

The first release of the DSSAT (V2.1, IBSNAT 1989) contained models of the following four crops: maize (CERES-Maize V2.1O), wheat (CERES-Wheat V2.1O;), soybean (SOYGRO V5.42) and groundnut (PNUTGRO V1.02).

The models for soybean, peanut, and dry bean were programmed as separate computer codes, and were referred to respectively as SOYGRO, PNUTGRO, and BEANGRO (HOOGENBOOM et al., 1992). Because of their similarity and the difficulty of maintaining separate codes for each, the authors amalgamated the models into a single set of computer codes (HOOGENBOOM et al., 1994, BOOTE et al., 1997) for use in DSSAT v3. These models are now referred to as CROPGRO-Soybean, CROPGRO-Peanut, and CROPGRO-Dry Bean, for soybean, peanut, and dry bean, respectively (HOOGENBOOM et al., 1994).

The models simulate the timing of phenological stages as affected by temperature and daylength. Dry matter production is computed by daily or hourly canopy photosynthesis models. Dry matter partitioning is based on source-sink relationships during early vegetative growth and during reproductive growth. Partitioning during the later vegetative phase is based on empirical functions that vary by crop. Crop-specific data files provide coefficients that characterize functional responses of each crop to its environment. Cultivar-specific data files provide coefficients to simulate the response of different cultivars to environment. For example, cultivar-specific coefficients quantify the photoperiod and temperature responsiveness of a cultivar as well as characteristics of vegetative and reproductive growth (TSUJI; HOOGENBOON; THORTON, 1998).

Battisti, Brender, Sentehas (2019) used the System for Agrotechnology Transfer (DSSAT), used by Jones et al. (2003), for soybean modelling in Brazil. After this procedure, they calibrated the model Brazilian cultivars, once used for Battisti et al. (2017b), with the model presenting a root mean square error below of 550 kg.ha^{-1} of soybean yield for calibration and validation phases. Then, they worked with reference evapotranspiration estimated by Penman-Monteith FAO 56 method (ALLEN et al. 1998); infiltration of water into the soil defined by soil curve number (SOIL CONSERVATION SERVICE 1972); soil water balance determined by Ritchie tipping-bucket method (RITCHIE, 1998); soil evaporation estimated by Suleiman-Ritchie (SULEIMAN; RITCHIE 2003); and leaf-level photosynthesis response from Boote and Pickering (1994).

Several authors applied the crop modelling simulations on their studies and they had satisfactory results and conclusions to help the farmer in decision-making (DALLACORT et al., 2008; BATTISTI; SENTELHAS, 2014; SINCLAIR et al., 2010; BOOTE, 2011; WHITE et al., 2011; NENDEL et al., 2014; DO RIO et al., 2015; SENTELHAS et al., 2015; BATTISTI, 2016; FISCHER, 2015; BATTISTI et al.,2017b; BATTISTI; BRENDER; SENTELHAS., 2019).

Therefore, crop modelling simulations are considered mechanisms of transforming information in useful knowledge to apply in agriculture, in addition, showing significant contribution to scientific and technology advances, aiming to understand the interactions between crop modelling simulations and the crop productions in the field (DOURADO NETO et al., 1998; SENTELHAS et al., 2015; BATTISTI, BENDER; SENTELHAS, 2019).

This simulation based on agrometeorological variables contributes to understand the biomass accumulation by plants in their different phenological stages, because consider the impacts from the weather conditions (ASSAD et al.,2007; POPP et al., 2003).

DSSAT program is a remarkable tool to use in experiments to contribute to science and in fields, helping the decision-making, the user, however, should be careful with the methodology procedure to obtain the necessary data (FERREIRA; DAPPER, 2023)

The models favour the simulation of many processes on soil and crop in order to provide information for the analysis of the sustainability of natural resources with increased the yield (BOWEN et al., 1993). According to the Oliveira et al. (2012), the application of simulation models saves time, work and amount of resources, bringing benefits for farmers and for the crops productions.

Zhang, Walker and Davis (2002) classified the models according to their complexity, from the simplest to the most complex, according to the amount of information to be entered. Because of the classification of the complexity of the models, errors occur related to each one of them: systematic errors, characteristic of simple models, and calibration errors resulting from more complex models. However, Palosuo et al. (2011) concluded there is no “program error” or even “best model”, but the wrong results come from the input data and interpretation.

Soybean grain demand is increasing constantly as a function of population growth; thus, it is necessary to improve production, mainly by yield increase, in a sustainable way (SENTELHAS et al., 2015). In this context, crop models can help in the assessment of the best strategies to achieve this goal, provided they are able to evaluate the best sowing dates (BATTISTI; SENTELHAS, 2014), the effects of climate change on yield, identify drought tolerance traits as well as many other possibilities (BATTISTI, 2016).

Different crop models have been developed and evaluated for soybean crop using data from specific regions and field scales. Beyond of parametrization from these regions, these crop models differ in the way and level that they simulate dynamic processes, as water balance and crop growth and development (WHITE et al., 2011), creating uncertainties related to the model´s parameters and structure (PALOSUO et al., 2011). It is important to highlight that the level of model complexity is not related with accuracy of results (SENTELHAS, 2011), once well calibrated simple models can be as accurate as the complex ones.

Complex crop models have more details in the description of all crop processes, which increase the number of possible uses of them to evaluate crop development and yield, and the influence of crop management on these variables (BATTISTI, 2016). For example, CROPGRO model integrates carbon, nitrogen and water balances on growth processes (BOOTE et al., 2003). These characteristics associated with other important processes, as the relationship between photosynthesis and [CO_2], make the models able to simulate soybean yield for several conditions, including under future climate scenarios (BOOTE et al., 1997; ALAGARSWAMY et al., 2006). Under this point of view, it is important to know the level of interaction between different processes on the crop models, identifying limiting points and possibilities of simulations with each one (BATTISTI, 2016).

Simulations models allow a better understanding of the components of crops productions and their limiting and reducing factors, as mentioned before in this Thesis, in addition, they make a better view of the components of productions agricultural system, in order to generate results to allow actions for agricultural planning, as the crops monitoring across the growing season and in several levels, such as the selection of cultivars, planting times, planning the use of agricultural machinery, among others (BATTISTI; BENDER; SENTELHAS, 2019).

According to Dourado Neto et al. (1998), analysis of statistics is one of the most used to adjust mathematical models or a curve to the experimental data. Being the simplest technique of modelling, it has been used since the beginning of the century. Adjusted models for regression have been used to foresee the effects on crop yield of management strategies, as application of fertilizers and variation of planting/sowing densities.

Thus, simulation models can show applications that consider from sowing to harvest, which helps in farmers and governmental institutions to decision-making. (RITCHIE 1998; HOOGENBOOM, 2000; GEDANKEN et al., 1998; BATTISTI; BENDER; SENTELHAS, 2019).

As mentioned, the crop simulation models can be used for many purposes, being an important tool to choose the best agricultural management strategies in different locations (WHITE et al., 2011). In addition, they can be applied to determine the best sowing or planting dates (DALLACORT et al., 2008; BATTISTI; SENTELHAS, 2015); the best cultivars with drought crops tolerance to use in experiment (BOOTE et al., 2003; LI et al., 2013; BAO et al., 2015; SINCLAIR et al., 2010; BOOTE, 2011) and the irrigation management in the field experiments (DOGAN et al., 2007).

The range of the applicability of modelling simulation is great, such as climate change scenarios and analysis of yield gap in soybean crop (MATTHEWS et al., 2013; EWERT et al., 2015; SENTELHAS et al., 2015). Comprehending the field experiments and their components of products productions and agricultural management, helps the researcher to observe correctly what the simulations models show (BATTISTI, 2016).

The aspects come from the plants are also relevant for crop modelling simulations. Breeding programs can introduce new cultivars with higher drought tolerance, for example (SINCLAIR et al., 2010) and the use of crop management strategies to reduce the yield gap, as use of implantation of irrigation system (NENDEL et al., 2014), adoption of new sowing dates (DO RIO et al., 2015), maturity groups (KUMAGAI; SAMESHIMA, 2014) and plant densities (VAZQUEZ et al., 2014).

In addition, an improved cultivar can be developed by adjusting and adapting the crop to the environment (GILBERT et al., 2011). For example, cultivars with more temperature and droughts resistance can be useful to reduce yield gap (BOOTE, 2011).

According to Sampaio et al. (2021) the soybean crop management is essential for models, such as crop simulation models oy soybean (CSM soybean), as association of irrigation or rainfed areas, plant population, plant density, and sowing dates and density (KUSS et al.,2008; NENDEL et al., 2014; SAMPAIO et al., 2021).

REFERENCES

AGOVINO, M. et al. Agriculture, climate change and sustainability: The case of EU-28. **Ecological Indicators**, Ecological Indicators, v. 105, p. 525-543, 2019.

ANDRADE, F.H., P. CALVINO, A. CIRILO, AND P. BARBIERI. Yield responses to narrow rows depend on increased radiation interception. **Agron. J**. 94:975–980, 2002.
doi:10.2134/agronj2002.0975

AGUIAR, E. B. **Produção e qualidade de mandioca de mesa (*Manihot esculenta* Crantz) em diferentes densidades populacionais e épocas de colheita.** 2003, 90 f. Dissertação (Mestrado em Agricultura Tropical e Subtropical) Instituto Agronômico de Campinas – IAC, Campinas, 2003.

ALBUQUERQUE, J. A. A.; SEDIYAMA, T.; SILVA, A. A.; CARNEIRO, J. E. S.; CECON, P. R.; ALVES, J. M. A. Interferência de plantas daninhas sobre a produtividade da mandioca (*Manihot esculenta*). **Planta Daninha**, Viçosa, v. 26, n. 2, p. 279-289, 2008.

ALLEN, R.; PEREIRA, L. S.; RAES, D.; SMITH, M. Crop evapotranspiration: guidelines for computing crop water requirements. **FAO Irrigation and Drainage Paper**, Rome, v. 56, p. 377-384, 1998.

ASANOME, N.;T. IKEDA. 2000. Effect of light interception difference in soybean canopy on leaf nitrogen concentration, nitrogen accumulation and nitrogen partitioning into pod. **Jpn. J. Crop. Sci**. 69:201–208, 2000. doi:10.1626/jcs.69.201

ASSAD, E. D.; MARIN, F. R.; EVANGELISTA, S. R.; PILLAU, F. G.; FARIAS, J. R. B. F.; PINTO, H. S.; ZULLO JUNIOR, J. Sistema de previsão da safra de soja para o Brasil. **Pesquisa Agropecuária Brasileira**, v. 42, n. 5, p. 615-625, 2007.

BAO, Y.; HOOGENBOOM, G.; MCCLENDON, R.W.; PAZ, J.O. Potential adaptation
strategies for rainfed soybean production in the south-eastern USA under climate change based on the CSM-CROPGRO-Soybean model. **Journal of Agricultural Science**, Cambridge, v. 153, p. 798-824, 2015.

BATTISTI, R.; SENTELHAS, P. C.; PILAU, F. G. Eficiência agrícola para as culturas da produção de soja, milho e trigo no Estado do Rio Grande do Sul entre 1980 e 2008. **Ciência Rural**, Santa Maria, v. 42, n. 1 p. 24-30, 2012.

BATTISTI, R.; SENTELHAS, P. C.; PILAU, F. G.; WOLLMANN, C. Eficiência climática para as culturas da soja e do trigo no Estado do Rio Grande do Sul em diferentes datas de semeadura. **Ciência Rural**, Santa Maria, v. 43, n. 3, p. 390-396, 2013.

BATTISTI, R.; SENTELHAS, P. Drought tolerance of Brazilian soybean cultivars simulated by a simple agrometeorological yield model. **Experimental Agriculture,** v. 51, n. 2, p. 285-298, 2015.

BATTISTI, R. **Calibration, uncertainties and use of soybean crop simulation models for evaluating strategies to mitigate the effects of climate change in Southern Brazil**. Doctorate Thesis in Agronomy (Agricultural Systems Engineering). University of São Paulo "Luiz de Queiroz" College of Agriculture. 189 f., 2016. Available in:https://www.teses.usp.br/teses/disponiveis/11/11152/tde-03102016-162340/en.php

BATTISTI, R.; SENTELHAS PC. Improvement of soybean resilience to drought through deep root system in Brazil. **Agronomy Journal** 109(4):1612– 1622, 2017. https://doi.org/10.2134/agronj2017.01.0023

BATTISTI. R.; SENTELHAS P.C.; BOOTE, K.J. Inter-comparison of performance of soybean crop simulation models and their ensemble in southern Brazil. **Field Crop Research** 200:28–37, 2017. https://doi.org/10.1016/j.fcr.2016.10.004

BATTISTI, R.; BENDER, F.D.; SENTELHAS, P.C Assessment of different gridded weather data for soybean yield simulations in Brazil. **Theoretical and Applied Climatology** 135, 237–247 (2019). https://doi.org/10.1007/s00704-018-2383-y

BHATIA, V.; SINGH, P.; WANI, S. P.; CHAUHAN, G. S.; RAO, A. V. R. K.; MISHRA, A. K.; SRINIVAS, K. Analysis of potential yields and yield gaps of rainfed soybean in India using CROPGRO-Soybean model. **Agricultural and forest meteorology,** v. 148, n. 8-9, p. 1252-1265, 2008.

BINDRABAN, P.; STOORVOGEL J. J.; JANSEN D. M.: VLAMING, J.; GROOT, J. J. R. Land quality indicators for sustainable land management: proposed method for yield gap and soil nutrient balance. **Agriculture, Ecosystems & Environment,** v. 81, n. 2, p. 103-112, 2000.

BOOTE, KJ., PICKERING, N.B. Modelling photosynthesis of row crop canopies. **Horticulture Science** v. 29:1423–1434, 1994.

BOOTE K J, JONES J W, HOOGENBOOM G, WILKERSON G G. Evaluation of the CROPGROSoybean model over a wide range of experiments. Pages 113-133 in Teng P S, Krop/f M J, 175 ten Berge H F M, Dent J B, Lansigan F P, van Laar H H (eds.) Applications of Systems Approaches at the Field Level, volume 2. **Kluwer Academic Publishers**, London, UK, 1997.

BOOTE, K.J.; JONES, J.W.; BATCHELOR, W.D.; NAFZIGER, E.D.; MYERS, O. Genetic coefficients in the CROPGRO-Soybean model: link to field performance and genomics. **Agronomy Journal**, Madison, v. 95, p. 32-51, 2003.

BOOTE, K.J. Improving soybean cultivars for adaptation to climate change and climate variability. In: YADAV, S.S.; REDDEN, R.J.; HATFIELD, J.L.; LOTZE-

CAMPEN, H.; HALL, A.E. **Crop adaptation to climate change**. West Sussex: Wiley-Blackwell, 2011. p. 370-395.

BOWEN, W. T.; JONES, J. W.; CARSKY, R. J. QUINTANA, J. O. Evaluation of the nitrogen submodel of Ceres-Maize following legume green manure incorporation. **Agronomy Journal**, Madison, v. 85, n. 1, p.153-159, 1993.

CAMARGO, A. P.; SENTELHAS, P. C. Avaliação do desempenho de diferentes métodos de estimativa da evapotranspiração potencial no Estado de São Paulo. **Revista Brasileira de Agrometeorologia**, Santa Maria, v. 5, n. 1, p. 89-97, 1997.

CALDANA et al. 2019. Agroclimatic Risk Zoning of Avocado (*Persea americana*) in the Hydrographic Basin of Paraná River III, Brazil. **Agriculture**, v.9, 2019 https://www.mdpi.com/2077-0472/9/12/263

CALDANA et al. 2021. Agroclimatic Risk Zoning for grape cultivation in the Basin of Paraná River III. **Amazonian Journal of Agricultural and Environmental Sciences**. v. 64, 2021

CALDANA et al. 2022. Analysis of precipitation an Indian Summer in Paraná state, Brazil – study of case in April 2021. **Revista Brasileira de Climatologia**. Dourados, v.31, Open Access, 24 p, 2022

CARAMORI, P. H. et al. Zoneamento agroclimático para o pessegueiro e a nectarineira no Estado do Paraná. **Revista Brasileira de Fruticultura**, v. 30, n. 4, p. 1040- 1044, 2008.

CARTER, O.G. Detailed yield analysis of the effect of different planting dates on seven soybean varieties. **Iowa State Journal**. 48:291-310, 1974.

CESB. Comitê Estratégico Soja Brasil, 2023. Disponível em: https://www.cesbrasil.org.br/

CINTRA, P.H.M. Variabilidade espacial e qualidade na semeadura de soja. **Braz. Ap. Sci. Rev**, Curitiba, v. 4, n. 3, p. 1206-1221 mai./jun. 2020

CUNHA, et al. Zoneamento agrícola e época de semeadura para soja no Rio Grande do Sul. **Revista Brasileira de Agrometeorologia**, Passo Fundo, v.9, n.3, (No Especial: Zoneamento Agrícola), p.446-459, 2001

DALLACORT, R.; FREITAS, P.S.L.; FARIA, R.T.; GONÇALVES, A.C.A.; REZENDE, R.; BERTONHA, A. Utilização do modelo Cropgro-soybean na determinação de melhores épocas de semeadura da cultura da soja, na região de Palotina, Estado do Paraná. **Acta Scientiarum Agronomy**, Maringá, v. 28, p. 583-589, 2008.

DOORENBOS, J.; KASSAM, A. H. Yield response to water. FAO Irrigation and Drainage Paper. n. 33. Rome: FAO, 1979. 193 p.

DOORENBOS, J.; KASSAM, A. H. **Efeito da água no rendimento das culturas**. Campina Grande: UFPB, 1994. 306 p.

DO RIO A, SENTELHAS, P.C.; FARIAS J.R.B.; SIBALDELLI RNR.; FERREIRA RC (2015). Alternative sowing dates as a mitigation measure to reduce climate change impacts on soybean yield in southern Brazil. **International Journal of Climatology** 36(11):3664–3672. https://doi.org/10.1002/joc.4583

DOURADO NETO, D.; TERUEL, D.A.; REICHARDT, K.; NIELSON, D.R.; FRIZZONE, J.A.; BACCHI, O.O.S. Principles of crop modelling and simulation: The implcations of the objective in model development. **Scientia Agricola**, v.55, p.51-57, 1998.

EGLI, D.B.; W.P. BRUENING. Potential of early-maturing soybean cultivars in late plantings. **Agron. J**. 92:532–537, 2000. doi:10.2134/agronj2000.923532x

EGLI, D.B.; CORNELIUS, P.L. A regional analysis of the response of soybean yield to planting date. **Agronomy Journal**, Madison, v. 101, p. 330-335, 2009.

EVANS, L.T., 1993. **Crop Evolution, Adaptation, and Yield**. Cambridge University Press, Cambridge, UK.

EMBRAPA. Empresa Brasileira de Pesquisa Agropecuária. CNPAF. Santo Antônio de Goiás, GO. **Zoneamento Agroclimático da Cultura do Arroz de Terras Altas para o Estado de Goiás.** Comunicado Técnico, 4p. 2008.

EWERT, F.; RÖTTER, R.P.; BINDI, M.; WEBBER, H.; TRNKA, M.; KERSEBAUM, K.C.; OLESEN, J.E.; VAN ITTERSUM, M.K.; JANSSEN, S.; RIVINGTON, M.; SEMENOV, M.A.; WALLACH, D.; PORTER, J.R.; STEWART, D.; VERHAGEN, J.; GAISER, T.; PALOSUO, T.; TAO, F.; NENDEL, C.; ROGGERO, P.P.; BARTOSOVÁ, L.; ASSENG, S. Crop modelling for integrated assessment of risk to food production from climate change. **Environmental Modelling & Software**, Oxford, v. 72, p. 287-303, 2015.

FAGUNDES, L. **Desenvolvimento, crescimento e produtividade da mandioca em função de datas de plantio.** 2009. 63 f. Dissertação (Mestrado em Agronomia) Universidade Federal de Santa Maria, Santa Maria, 2009.

FAO. FAO statistical yearbook 2013 world food and agriculture. Rome: Food and Agriculture Organization of the United Nations, 2013. 307 p.

FERREIRA, L.G.B. **Disponibilidade hídrica e produtividade de soja no Oeste do Paraná**. Dissertação de Mestrado, Instituto de Desenvolvimento Rural do Paraná. Londrina, 77 f., 2017

FERREIRA, L.G.B.; CALDANA, N.F.S.; MARTELÓCIO, A.C.; COSTA, A.B.F.; NITSCHE, P.R.; CARAMORI, P.H. Rainfall variability and analysis of droughts periods risks during the soybean crop (*Glycine max* L.) in the Western of Paraná State, Brazil. **Brazilian Journal of Climatology** v.27, p.590-611, 2020.

FIALHO, J. de F.; OLIVEIRA, M. A. S.; ALVES, R. T.; PEREIRA, A. V.; JUNQUEIRA, N. T. V.; GOMES, A. C. **Danos do Percevejo-de-Renda na Produtividade da Mandioca no Distrito Federal**. Comunicado técnico - Embrapa Cerrados, n. 48, p. 1-3, 2001.

FISCHER, R. A. Definitions and determination of crop yield, yield gaps, and of rates of change. **Field Crops Research,** v. 182, p. 9-18, 2015.

FREITAS, M de C, M. A cultura da soja no brasil: o crescimento da produção brasileira e o surgimento de uma nova fronteira agrícola. **Enciclopédia Biosfera**, Centro Científico Conhecer - Goiânia, vol.7, N.12; 2011

GABRIEL, L. F.; STRECK, N. A.; UHLMANN, L. O.; SILVA, M. R.; SILVA, S. D. Mudança climática e seus efeitos na cultura da mandioca. **Revista Brasileira de Engenharia Agrícola e Ambiental**, Campina Grande, v. 18, n. 1, p. 90-98, 2014.

GEDANKEN, A.; MANTOVANI, E. C.; MANTOVANI, B. H. M.; COSTA, L. C.; SANS, L. M. A.; FREITAS, P. S. L. de. Utilização do modelo CERES-Maize para avaliar estratégias de irrigação em duas regiões de Minas Gerais. **Acta Scientiarum. Agronomy**, Maringá, v. 25, n. 2, p. 439-447, 2003.

GETZ, W.M.; GUTIERREZ, A.P. A perspective on systems analysis in crop production and insect pest management. **Annual Review of Entomology**, v.27, p.447, 1982.

GILBERT, M.E.; HOLBROOK, N.M.; ZWIENIECKI, M.A.; SADOK, W.; SINCLAIR, T.R.
Field confirmation of genetic variation in soybean transpiration response to vapor pressure deficit and photosynthetic compensation. **Field Crop Research**, Amsterdam, v. 124, p. 85-92, 2011.

GODFRAY, H. C. J.; BEDDINGTON, J. R.; CRUTE, I. R.; HADDAD, L.; LAWRENCE, D.; MUIR, J. F.; PRETTY, J.; ROBINSON, S.; THOMAS, S. M.; TOULMIN, C. Food security: the challenge of feeding 9 billion people. **Science,** v. 327, p. 812-818, 2010.

HEATHERLY, L.G. 2005. Soybean development in the midsouthern USA related to date of planting and maturity classifi cation. www.plantmanagementnetwork.org/cm/.
Crop Manage.:doi: 10.1094/CM-2005–0421–01-RS.

HEINEMANN, A. B.; SENTELHAS, P. C. Environmental group identification for upland rice production in central Brazil. **Scientia Agricola,** v. 68, n. 5, p. 540-547, 2011.

HEINEMANN, A.B.; RAMIREZ-VILLEGAS, J.; SOUZA TLPO, DIDONET, A.D.; DI STEFANO, J.G.; BOOTE, K.J.; JARVIS A. Drought impact on rainfed common bean production areas in Brazil. **Agricultural and Forest Meteorology** v.225: 57–74, 2016
https://doi.org/10.1016/j.agrformet.2016.05.010

HERRERA CAMPO, B. V.; HYMAN, G.; BELLOTTI, A. Threats to cassava production: known and potential geographic distribution of four key biotic constraints. **Food Security**, p. 1-17, 2011.

HOOGENBOOM G, JONES J W, BOOTE K J. Modeling growth, development and yield of grain legumes using SOYGRO, PNUTGRO, and BEANGRO: A review. **Trans ASAE** 35(6)2043-2056, 1992.

HOOGENBOOM G, JONES J W, WILKENS P W, BATCHELOR W D, BOWEN W T, HUNT L A, Pickering N B, Singh U, Godwin D C, Baer B, Boote K J, Ritchie J T, White J W. Crop models. Pages 94-244 in Tsuji G Y, Uehara G, Balas S (Eds.) DSSAT v3 Vol 2-2. University of Hawaii, Honolulu, HI, 1994

HOOGENBOOM, G. Contibution of agrometeorology to the simulation os crop production and its applications. **Agricultural and Forest Meteorology**, n. 103, p. 137-157, 2000.

HU, M.; WIATRAK, P. Effect of Planting Date on Soybean Growth, Yield, and Grain Quality: Review. **Agronomy Journal**. vol.104, 2012.

HUNT, L.A.; BOOTE, K.J. Data for model operation, calibration, and evaluation. In: TSUJI, G.Y.; HOOGENBOOM, G.; THORNTON, K. **Understanding options for agricultural production**. Dordrecht: Kluwer Academic, 1998.

JONES, J.W.; HOOGENBOOM, G.; PORTER, C.H.; BOOTE,K.J.; BATCHELOR, W.D, HUNT, L.A.; WILKENS, W.; SINGH, U.; GIJSMAN AJ.; RITCHIE, J.T. The DSSAT cropping system model. **European Journal of Agronomy** 18(3-4):234–265, 2003 https://doi.org/10.1016/S1161-0301(02)00107-7

JOYCE, L.A.; KICKERT, R.N. Applied plant growth models for grazinglands, forests and crops. In: WISIOL, K.; HESKETH, J.D. Plant growth modeling for resource management: current models and methods. Boca Raton: CRC Press.v.1, p.141-156, 1987.

KUMAGAI, E.; SAMESHIMA, R. Genotypic differences in soybean yield responses to
increasing temperature in a cool climate are related to maturity group. **Agricultural and Forest Meteorology**, Amsterdam, v. 198/199, p. 265-272, 2014.

KUSS, R.C.R.; KÖING, O.; DUTRA, L.M.C.; BELLÉ, R.A.; ROGGIA, S.; STURMER,
G.R. Populações de plantas e estratégias de manejo de irrigação na cultura da soja. **Ciência Rural,** Santa Maria, v. 38, p. 1133-1137, 2008.

LAHAI, M. T.; GEORGE, J. B.; EKANAYAKE, I. J. Cassava (*Manihot esculenta* Crantz) growth indices, root yield and its components in upland and inland valley ecologies of Sierra Leone. **Journal of Agronomy and Crop Science**, v. 182, n. 4, p. 239-248, 1999.

LEHMANN, N.; FINGER, R.; KLEIN, T.; CALANCA, P.; WALTER, A. Adapting crop management practices to climate change: modeling optimal solutions at the field scale.
Agricultural Systems, Essex, v. 117, p. 55-65, 2013.

LIU, Z.; YANG, X.; LIN, X.; HUBBARD, K.G.; LV, S.; WANG, J. Maize yield gaps caused by non-controllable, agronomic, and socioeconomic factors in a changing climate of Northeast China. **Science of the Total Environment** v. 541, p. 756–764, 2016.

LOBELL, D. B., CASSMAN, K. G.; FIELD, C. B. Crop yield gaps: their importance, magnitudes, and causes. **Annual Review of Environment and Resources**, v. 34, p. 179-204, 2009.

MARTINS, M. A. **Estimativa de produtividade das culturas do milho e do sorgo a partir de modelos agrometeorológicos em algumas localidades da região nordeste do Brasil**. 2012. 97 f. Dissertação (Mestrado em Meteorologia) Instituto Nacional de Pesquisa Espaciais – INPE, São José dos Campos, 2012.

MARTINS, J. A. et al. Climatology of destructive hailstorms in Brazil. **Atmospheric Research**, v. 184, p. 126-138, 2017.

MATTHEWS, R.B., RIVINGTON, M., MUHAMMED, S., NEWTON, A.C., HALLETT,
P.D. Adapting crops and cropping systems to future climates to ensure food security: the role of crop modelling. **Global Food Security**, Amsterdam, v. 2, p. 24-28, 2013.

MATTOS, P. D.; CARDOSO, E. Cultivo da Mandioca para o Estado do Pará. **Cultivo da mandioca para o Estado do Pará,** v. 13, 2003.

MEOTTI, G.V.; BENIN, G.; SILVA, R.R.; BECHE, E.; MUNARO, L.B. Épocas de semeadura e desempenho agronômico de cultivares de soja. **Pesquisa Agropecuária Brasileira**, v.4, p.14-21, 2012.

MONTEITH, J.L. Solar-radiation and productivity in tropical ecosystems. **J. Appl. Ecol**. 9:747–766., 1972 doi:10.2307/2401901

MORAES, A. V. de C.; CAMARGO, M. B. P. de; MASCARENHAS, H. A. A.; MIRANDA, M. A. C. de; PEREIRA, J. C. V. N. A. Teste e análise de modelos agrometeorológicos de estimativa de produtividade para a cultura da soja na região de Ribeirão Preto. **Bragantia**, Campinas, v. 57, n. 2, 1998.

MORETO, V. B.; ROLIM, G. S. Estimation of annual yield and quality of 'Valência' orange related to monthly water deficiency. **African Journal of Agricultural Research**, v. 10, n. 6, p. 543-553, 2015.

NAKAGAWA, J.; ROSOLEM, C.A.; MACHADO, J.R. Épocas de semeadura de soja: Efeitos na produção de grãos e nos componentes de produção. **Pesq. agropec. bras.**, Brasilia, 18(l1):1187-1198,nov. 1983

NENDEL, C.; KERSEBAUM, K.C.; MIRSCHELW, WENKEL, K.O management options as climate change adaptation strategies using the MONICA model. **European Journal of Agronomy** 52:47–56, 2014 https://doi.org/10.1016/j.eja.2012.09.005

NITSCHE, P. R. et al. **Atlas Climático do Estado do Paraná.** 2019. Disponível em: < https://www.idrparana.pr.gov.br/Pagina/Atlas-Climatico>

OLIVEIRA, S. L.; MACÊDO, M. M. C.; PORTO, M. C. M. Efeito do déficit de água na produção de raízes de mandioca. **Pesquisa Agropecuária Brasileira**, Brasília, v. 17, n. 1, p. 121-124, 1982.

OLIVEIRA, R. A.; SANTOS, R. S. dos; RIBEIRO, A.; ZOLNIER, S.; BARBOSA, M. H. P. Estimativa da produtividade da cana-de-açúcar para as principais regiões produtoras de Minas Gerais usando-se o método ZAE. **Revista Brasileira de Engenharia Agrícola Ambiental**, v. 16, n. 5, p. 549–557, 2012.

ONU. United nations, department of economic and social affairs. The United Nations, Population Division, Population Estimates and Projections Section, 2021.

PALOSUO, T.; KERSEBAUM, K.C.; ANGULO, C.; HLAVINKA, P.; MORIONDO, M.; OLESEN, J.E.; PATIL, R.H.; RUGET, F.; RUMBAUR, C.; TAKÁC, J.; TRNKA, M.; BINDI, M.; CALDAG, C.; EWERT, F.; FERRISE, R.; MIRSCHEL, W.; SAYLAN, L.; SISKA, B.; RÖTTER, R. Simulation of winter wheat yield and its variability in different climates of Europe: a comparison of eight crop growth models. **European Journal of Agronomy**, Amsterdam, v. 35, p. 103-114, 2011.

PATHMESWARAN, C. et al. Impact of extreme weather events on coconut productivity in three climatic zones of Sri Lanka. **European Journal of Agronomy**, v. 96, p. 47-53, 2018.

PEART, R.M.; BARRET JR., J.R. Simulation in crop ecosystem management. In: **Winter simulation conference. simulation councils**, La Jolla. Proceedings. p.389, 1976.

PEDERSEN, P.; J.G. LAUER. 2003. Soybean agronomic response to management systems in the upper Midwest. **Agron. J**. 95:1146–1151. doi:10.2134/agronj2003.1146

PEREIRA, A. R.; ANGELOCCI, L. R.; SENTELHAS, P. C. **Agrometeorologia: Fundamentos e Aplicações Práticas**. Guaíba: Ed. Agropecuária, 2002. 478p.

PONTI, T.; RIJK, B.; VAN ITTERSUM, M. K. The crop yield gap between organic and conventional agriculture. **Agricultural Systems,** v. 108, p. 1-9, 2012.

POPP, M. P.; DILLON, C. R.; KEISLING, T. C. Economic and weather influences on soybean planting strategies on heavy soils. **Agricultural Systems**, v. 76, n. 3, p. 969-984, 2003.

PRADO, I. N. Avaliação produtiva e econômica da substituição do milho por subprodutos industriais da mandioca na terminação de novilhas. **Campo Digital**, Campo Mourão, v. 1, n. 1, p. 37-47, 2006.

PRIESTLEY, C.H.B.; TAYLOR, R. J. On the assessment of surface heart flux and evaporation using large-scale parameters. **Monthly Weather Review** v.100, 81–92, 1972

PURCELL, L.C., R.A. BALL, J.D. REAPER, AND E.D. VORIES. Radiation use efficiency and biomass production in soybean at different plant population densities. **Crop Sci**. 42:172–177, 2002. doi:10.2135/cropsci2002.0172

RAHMAN, M.M., J.G. HAMPTON, M.J. HILL. 2005. The effect of time of sowing on soybean seed quality. **Seed Sci. Technol**. 33:687–697.

REYNOLDS, J. F. Some misconceptions of mathematical modeling. **What's New Plant Physiology**, v. 10, n. 11, p. 41-44, 1979.

RICCE, W. da S. et al. Zoneamento agroclimático da cultura do abacaxizeiro no Estado do Paraná. **Semina: Ciências Agrárias**, v. 35, n. 4, p. 2337-2346, 2013.

RITCHIE, J.T. **Soil water balance and plant water stress**. In: Tsuji GY, Hoogenboom G, Thornton K (eds) Understanding options of agricultural production. Kluwer Academic Publishers and International Consortium for Agricultural Systems Applications, Dordrecht, The Netherlands, pp 41–53, 1998. https://doi.org/10.1007/978-94-017-3624-4_3

SAATH, K. C. de O.; FACHINELLO, A. L. Crescimento da Demanda Mundial de Alimentos e Restrições do Fator Terra no Brasil. **Revista de Economia e Sociologia Rural**, Brasília, v. 56, n. 2, p. 195-212, 2014.

SAKAMOTO, C.; Le DUC, S. Sense and nonsense: statistical crop growth and yield models. In: WEISS, A. (Ed.) **Computer techniques and meteorological data applied to problems of agriculture and forestry**: workshop. Boston: American Meteorological Society, 1981. 185p.

SANTI, A. et al. Impacto de cenários futuros de clima no zoneamento agroclimático do trigo na região Sul do Brasil. **Agrometeoros**, v. 25, n. 2, p. 303-311, 2018.

SAMPAIO, L.S.; BATTISTI, R.; LANA, M.A.; BOOTE, J.K. Assessment of sowing dates and plant densities using CSMCROPGRO-Soybean for soybean

maturity groups in low latitude. **The Journal of Agricultural Science** 1–14, 2021
https://doi.org/
10.1017/S0021859621000204

SANTOS, T.G.; BATTISTI, R.; CASAROLI, D.; ALVES JÚNIOR, J.; EVANGELISTA, A.W.P. Assessment of agricultural efficiency and yield gap for soybean in the Brazilian Central Cerrado biome. **Bragantia**, Campinas, 80, 2021
https://doi.org/10.1590/1678-4499.20200352

SCHIMITT, A. T. Principais insetos pragas da mandioca e seu controle. In: CEREDA, M. P. **Agricultura: tuberosas amiláceas latino-americanas**. São Paulo: Fundação Cargill, 2002. p. 350-369.

SENTELHAS, P.; BATTISTI, R.; CÂMARA, G. M. S.; FARIAS, J. R. B.; HAMPF, A. C.; NENDEL, C. The soybean yield gap in Brazil–magnitude, causes and possible solutions for sustainable production. **The Journal of Agricultural Science**, p. 1-18, 2015.

SENTELHAS, P. C.; BATTISTI, R.; MONTEIRO, L. A.; DUARTE, Y. C. N.; VISSES, F. A. Yield gap: conceitos, definições e exemplos. **Informações Agronômicas**, Piracicaba, International Plant Nutrition Institute, n. 155, p. 9-12, 2016. Disponível em: <http://www.ipni.net/publication/ia-brasil.nsf/0/9C05063FB033C24A83258042004C85 94/>. Acesso em: 28 de set. de 2020.

SILVESTRE, M. R. et al. Critérios estatísticos para definir anos padrão: uma contribuição à climatologia geográfica. **Revista Formação**, v. 2, n. 20, p. 23-53, 2013.

SINCLAIR, T.R.; MESSINA, C.D.; BEATTY, A.; SAMPLES, M. Assessment across the
United States of the benefits of altered soybean drought traits. **Agronomy Journal,** Madison, v. 102, p. 475-482, 2010.

SOIL CONSERVATION SERVICE (SCS). **National engineering handbook**. Hydrology Section 4, Chapters 4/10, 1972.

SOUZA, R. C. de. **Avaliação do potencial agronômico de cultivares de mandioca oriundas do nordeste brasileiro.** 2017, 31 f. Dissertação (Mestrado Profissional em Olericultura) Instituto Federal de Educação, Ciência e Tecnologia Goiano, Morrinhos, 2017.

SOUZA, D. C. F. et al. Zoneamento Agroclimático da Palma Forrageira (Opuntia Sp) Para o Estado de Sergipe. **Revista Brasileira de Agricultura Irrigada**, v. 12, n. 1, p. 2338, 2018.

SPEHAR, C.R; FRANCISCO, E.R AND PEREIRA, E.A (2015) Yield stability of soybean cultivars in response to sowing date in the lower latitude Brazilian Savannah Highlands. **Journal of Agricultural Science** 153, 1059–1068.

STRECK, N. A.; PINHEIRO, D. G.; ZANON, A. J.; GABRIEL, L. F.; ROCHA, T. S. M.; SOUZA, A. T.; da SILVA, M. R. Efeito do espaçamento de plantio no crescimento, desenvolvimento e produtividade da mandioca em ambiente subtropical. **Bragantia**, Campinas, v. 73, n. 4, p. 407-415, 2014.

SULEIMAN, A.A.; RITCHIE, J.T. Modelling soil water redistribution during second-stage evaporation. **Soil Science Society of American Journal**. v. 67(2):377–386, 2003. https://doi.org/10.2136/sssaj2003.3770

TAKATSU, A.; FUKUDA, C. Current status of cassava diseases in Brazil. In: **Integrated pest management for tropical root and tuber crops: proceedings of the workshop on the global status of and prospects for integrated pest management of root and tuber crops in the tropics held in Ibadan, Nigeria**. Ibadan, Nigéria: International Institute of Tropical Agriculture, 1990. p. 127-134.

TAYT'SOHN, F. C. O. Assessing sugarcane expansion to ethanol production under climate change scenarios in Paranaíba river basin–Brazil. **Biomass and Bioenergy**, v. 119, p. 436-445, 2018.

TEIXEIRA, W.W.R.; BATTISTI, R.; SENTELHAS, P.C.; MORAES, M. F de.; OLIVEIRA JÚNIOR. Uncertainty assessment of soya bean yield gaps using DSSATCSM- CROPGRO-Soybean calibrated by cultivar maturity groups. **Journal of Agronomy and Crop Science** (Wiley), n.205, p.533-544, 2018.

TEIXEIRA, W.W.R; BATTISTI, R; SENTELHAS, P.C, DE MORAES MF and de OLIVEIRA JUNIOR A. Uncertainty assessment of soya bean yield gaps using DSSAT-CSM-CROPGRO-soybean calibrated by cultivar maturity groups. **Journal of Agronomy and Crop Science** 205, 1–12, 2019.

THORNTHWAITE, C. W.; MATHER, J. R. The water balance. Centerton, USA: Drexel Institute of Technology - Laboratory of Climatology, 1955. 104 p. (Publications in Climatology, vol. VIII, n.1).

TREMACOLDI, C. R. Manejo das principais doenças da cultura da mandioca no estado do Pará. In: MODESTO JUNIOR, M. S.; ALVES, R. N. B. **Cultura da mandioca: aspectos socioeconômicos, melhoramento genético, sistemas de cultivo, manejo de pragas e doenças e agroindústria**. 1. ed. Brasília: Embrapa, 2016. cap. 9. p. 162-170.

TSUJI, G.Y.; HOOGENBOON, G.; THORTON, P.K. **Understanding options for agricultural production**. Springer Science, 404 p., 1998

USDA – UNITED STATES DEPARTAMENT OF AGRICULTURE: International Production Assessment Division. Brazilian soybean productions, 2023 <https://ipad.fas.usda.gov/countrysummary/Default.aspx?id=BR&crop=Soybean >

VALLE, T. L.; CARVALHO, C. R. L.; RAMOS, M. T. B.; MÜHLEN, G. S.; VILLELA, O. V. Cyanide and acid content in progenies from crosses of bitter and sweet cassava cultivars. **Bragantia**, Campinas, v. 63, n. 2, p. 221-226, 2004.

VALLE, T. L.; FELTRAN, J. C.; CARVALHO, C. R. L. **Mandioca para produção de etanol**. Disponível em: < https://www.agencia.cnptia.embrapa.br/Repositorio/ mandiocaetanol2_000g7fa3wbe0 2wx5ok0wtedt3i5fuc84.pdf >. Acesso em: 20 set. 2020.

VAN ITTERSUM, MARTIN K.; CASSMANB, K. G.; GRASSINI, P. B.; JOOST W. A.; TITTONELLI, P. C.; HOCHMAND, Z. Yield gap analysis with local to global relevance - a review. **Field Crops Research**, v. 143, p. 4-17, 2013.

VAN ROEKEL RJ, PURCELL LC AND SALMERÓN M (2015) Physiological and management factors contributing to soybean potential yield. **Field Crops Research** 182, 86–97.

VAZQUEZ, G.H.; PERES, A.R.; TARSITANO, M.A.A. Redução na população de plantas de soja e o retorno econômico na produção de grãos. **Cientifica**, Jaboticabal, v. 42, p. 108-117, 2014.

WALLACH, D.; MAKOWSKI, D.; JONES, J.W. **Working with dynamic crop models:**
evaluation, analysis, parameterization, and application. Amsterdam: Elsevier, 2006. 447 p.

WEAVER, D.B., R.L. AKRIDGE, AND C.A. THOMAS. Growth habit, planting date, and row spacing effects on late-planted soybean. **Crop Sci**. 31:805–810. 1991 doi:10.2135/ cropsci1991.0011183X003100030052x

WILLMOTT, C. J., ACKLESON, S. G., DAVIS, R. E., FEDDEMA, J. J., KLINK, K. M., LEGATES, D. R., O'DONNELL, J.; ROWE, C.M.. Statistics for the evaluation and comparison of models. **Journal of Geophysical Research**: v. 90, p. 8995–9005, 1985

WHITE, J.W.; HOOGENBOOM, G.; KIMBALL, B.A.; WALL, G.W. Methodologies for simulating impacts of climate change on crop production. **Field Crops Research**, Amsterdam, v. 124, p. 357-368, 2011.

WIRÉHN, L. Nordic agriculture under climate change: A systematic review of challenges, opportunities and adaptation strategies for crop production. **Land use policy**, v. 77, p. 63-74, 2018

ZANON, A.J.; STRECKNA, GRASSINI P. Climate and management factors influence soybean yield potential in a subtropical environment.
Agronomy Journal 104(4):1447–1454, 2016.
https://doi.org/10.2134/agronj2015.0535

ZDZIARSKI, AD; TODESCHINI, M.H; MILIOLI AS; WOYANN LG; MADUREIRA A, STOCO MG AND BENIN G (2018) Key soybean maturity groups to increase grain yield in Brazil. **Crop Science** 58, 1155–1165.

ZHANG, L.; WALKER, G. R.; DAWES, W. R. Water balance modeling: Concepts and applications. In: MEVICAR, T. R.; RUI, L.; WALKER, J.; FITZPATRICK, R. W.; CHANGMING, L. Regional Water and Soil Assessment for Managing Sustainable Agriculture in China and Australia. **ACIAR**, Canberra, p. 31-47, 2002.

ZHANG, Q.Y., Q.L. GAO, S.J. HERBERT, Y.S. LI, AND A.M. HASHEMI. Influence of sowing date on phenological stages, seed growth and marketable yield of four vegetable soybean cultivars in North-eastern USA. **African J. Agric. Res**. 5:2556–2562, 2010

WATER AVAILABILITY AND RISK OF DRY SPELLS FOR SOYBEAN IN THE WEST OF PARANÁ STATE, BRAZIL

ABSTRACT

The water availability is the factor that most affects the production of soybean crops in the state of Paraná. Dry spells are extremely damaging to agricultural production and, therefore, studies that identify the frequency and intensity of these events are relevant to the planning of operations. Thus, the objective of this study was to evaluate the rainfall availability and to determine the frequency of dry spells in the Meso-region of Parana state. Precipitation data were collected from 48 stations distributed throughout the mesoregion between 1976 and 2018. Rainfall variability was analyzed in the annual, monthly and 10-days scales, and the frequencies of dry periods of 10 days from September to March and ≥ 20 days during the year. It was verified that the mesoregion showed differences in the amount of precipitation, but with enough rainfall for soybean cultivation. The relief favors the distribution of rainfall in the region. The southern portion has the highest mean precipitation, while the northern portion has the lowest. The dry spells frequencies per moving 10-day periods showed a maximum of 35% between the months of September and March in the region. The periods with the lowest risk are throughout the month of October, followed by December 10 to January 5. The biggest risks of dry spells are concentrated in the month of September, due to the irregular start of the rainy season. The results of this study help to choose the best sowing times such that the most sensitive phases of soybean cycle occur in the lowest risk periods.

Key-words: agricultural planning; rainfall variability; dry periods, climatic risk.

INTRODUCTION

The most significant natural variables in the production process, regardless of the mode of production come from the climate and are considered inputs and the main source of energy in the terrestrial system (SANT'ANNA NETO, 2015). Despite the availability of advanced technological and scientific resources, climate variability remains a challenge and is the main cause of oscillations in agricultural productivity (CARAMORI et al., 2016).

The Western of Paraná State has a Cfa climate according to Koppen classification ((NITSCHE et al., 2019), with irregularities in the distribution of rainfall that cause frequent decreases in crop yields (MINUZZI; CARAMORI, 2015; JACONDINO et al., 2018). Water affects all the physiological processes that occur in the plant, such as root absorption, nutrient transport, thermoregulation and hydration, essential to maintain the structure and cellular activity (TAIZ AND ZEIGER, 2009). Several studies have focused on water deficit in the production of soybean crops. This could help comprehend the deficit water impacts in the soybean crops and offer alternatives to overcome or even cope with this situation (GRASSINI et al., 2015; SENTELHAS et al., 2015; DIETZEL et al., 2016; BATTISTI et al., 2017; GIMÉNEZ et al., 2017; AHMAD et al., 2018; DU et al., 2018; GAJIĆ et al., 2018; JHA et al., 2018; KHOJELY et al., 2018; ADEBOYE et al., 2019; AKHTAR et al., 2019; BENCKE-MALATO et al., 2019; JUNIOR; SENTELHAS, 2019).

Agrometeorological conditions directly interfere in the choice of crops to be produced in a given region, sowing time, use of agrochemicals, management practices and irrigation. Studies that identify and predict periods of drought, heavy rain, wind and frost contribute to crop planning and management, aiming at obtaining the best net income (BERLATO et al., 2005; LUNGARSKA; CHAKIR, 2018; PATHMESWARAN et al. 2018; KARIMI et al., 2018; WIRÉHN, 2018; AGOVINO et al., 2018; DOS REIS et al. 2017; ALBUQUERQUE et al., 2018; CONCEIÇÃO et al., 2018; GELCER et al. 2018; DE MATOS et al., 2018; MEDEIROS et al., 2018; SANTI et al., 2018; DE SOUSA; DE OLIVEIRA, 2018; DE SOUZA et al., 2018; TAYT'SOHN et al., 2018).

The soybean crop is the main agricultural activity in Brazil and in the state of Paraná, southern Brazil. It is cultivated in the Spring-Summer season, between the months of September and March, using early cultivars, to allow a second season cropping of maize. Rainfall variability and water deficit is the main cause of yield loss of soybean (FARIAS et al., 1997; PEDERSEN; LAUER, 2004; STÜLP et al., 2010; ALVARES et al., 2013; GARCIA et al., 2018; ZARO et al., 2018; JUNIOR; SENTELHAS, 2019).

It is clear that water availability is one of the most important elements for soybean crops, affecting the productive potential of the crop and causing variability of production, in Brazil (FARIAS et al., 2007, SENTELHAS et al., 2015, BATTISTI et al., 2017; GAJIĆ et al., 2018; BENCKE-MALATO et al., 2019).

The purpose of this work was to evaluate water availability and determine the frequency of dry periods during the Spring-Summer seasons, in order to contribute soybean crop planning in the Western of Paraná State, Brazil.

MATERIAL AND METHODS

Characterization of the study area

The Western of Paraná State has a population of about 1 million inhabitants (IBGE, 2018), where agriculture is still the main economic activity. In 2014, 79% of its area was occupied with agriculture, 62% of which was temporary / permanent crops and 17% was pasture. Soil occupation occurred rapidly, from 1950 to 1980 the area increased sixfold, and in this process, the temporary crops (maize, beans, rice and cassava) were replaced, by wheat, soybeans and corn (MEDEIROS; ZANÃO JUNIOR, 2015).

Soybean production in the Western of Paraná State was just over 4.0 million tons in the 2016/17 harvest, while in the other regions of the State, in the same crops cycle, production was approximately 17 million tons. The municipalities of Cascavel and Toledo stand out with annual production around 200,000 tons (Figure 1). This region produces 23.5% of all agricultural production in Paraná State.

Figure 1. Soybean production in the West of Paraná State, Brazil.

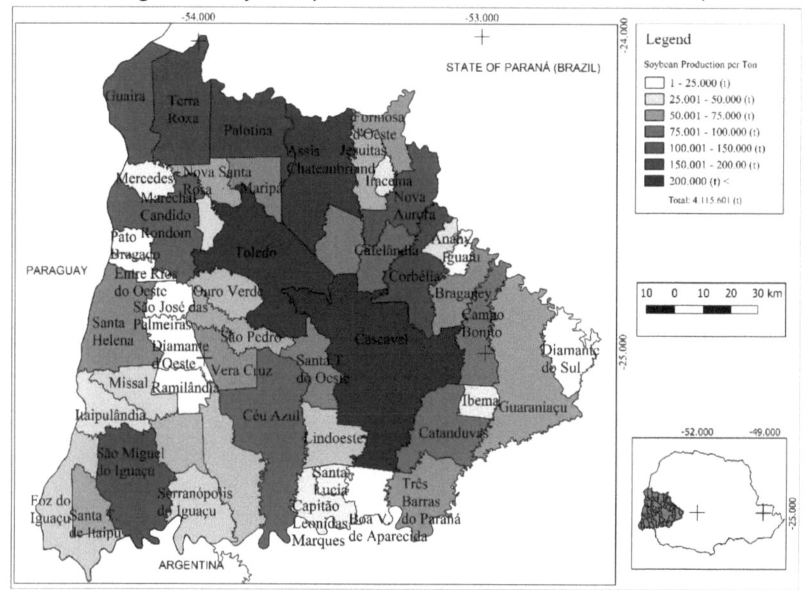

Source: Deral (2019); adapted and organized by the authors (2019).

Methodological Procedures

The database comprised daily pluviometric data from 48 stations from the networks of Instituto Agronômico do Paraná (IAPAR), Instituto das Águas Paraná (ÁGUAS PARANÁ), Instituto Nacional de Meteorologia (INMET), Agência Nacional das Águas (ANA) and Sistema Meteorológico do Paraná (SIMEPAR). The stations are distributed in West of Paraná State (Figure 2), with homogeneous periods of daily data observation from 1976 to 2018. Altimetric data were also used to identify possible geographic factors that may interfere with regional rainfall distribution. The relief of the region ascends from the North, South and West, to the Center-East direction with altitudes around 100-200 m to 800-900 m.

Figure 2. Hypsometry and location of the Meteorological Stations in the study area.

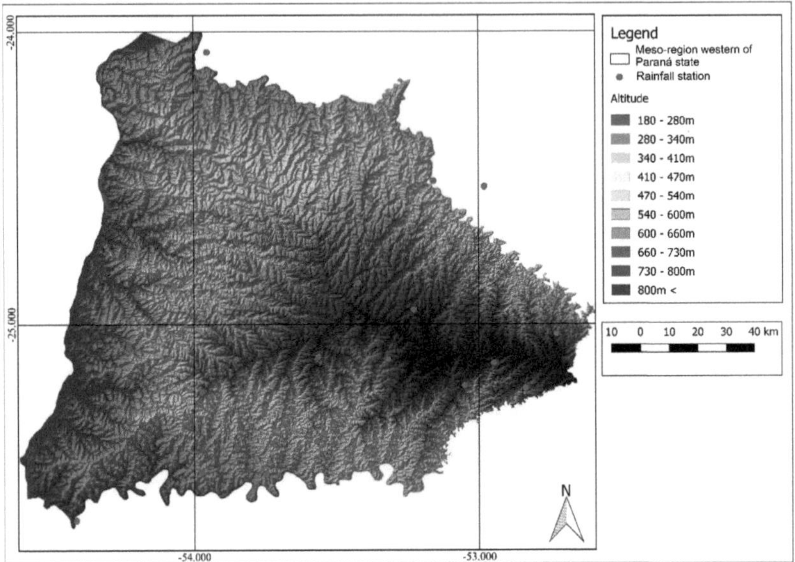

Source: ANA; ÁGUAS PARANÁ; INMET; IAPAR.

The spatial rainfall data of the basin were represented by interpolation using the *Inverse distance weighted* (IDW) algorithm, which is an appropriate method for the spatial representation of rainfall data (MUELER, 2004). The maps were created using the Qgis *software*. In this procedure, the maps were transformed into *raster format*, generating a fine mesh of grid cells (pixels) that incorporate a numerical value. Each pixel of the generated images has a spatial resolution of 1 km x 1 km (ELY; DUBREUIL, 2017).

The rainfall variability was explored through *Box Plot* or box diagram charts. This procedure is useful to provide a quick view of the data distribution. If the distribution is symmetrical the box is balanced with median equals to the mean and positioning in the center. For asymmetric distributions there is an imbalance in the box in relation to the median (SILVESTRE et al., 2014, p.27). The graphics were created with the program *Statistica® software*. The Box plots represent five values of classification. Outliers were divided into discrepant (values above the maximum considered but not extreme) and extremes, considered as any values greater than Q3 + 1.5 (Q3 - Q1) or lower than Q1 – 1.5 (Q3 - Q1). The maximum and minimum are considered the highest values of the series, but they are not extreme or discrepant. In the box, three quartiles with 25 % of the data are classified, in addition to the median value, equivalent to the second quartile, or 50 % of the data (LEM et al., 2013, SCHNEIDER; DA SILVA, 2014).

The analysis of Box plot was applied to the precipitation data of the stations Matelândia, Santa Lucia, Terra Roxa and Toledo (Figure 4).

Analysis of dry periods

This analysis consisted on the verification of the probabilities of consecutive dry days as consequence of the regional variability and distribution of rainfall. Precipitation events were considered as those with at least 1 mm. The following methods for the identification of risk of occurrence of dry spells were analyzed:

a) Consecutive periods of 10 days without rain: The soybean is grown in the Paraná State in the period from September to March, with differences in cycle due to sowing date and temperature. Dry spells during this period, known as "veranicos", were identified in a sequence of ten days without rainfall during the crop cycle. The assessments consisted on determining the frequencies of the number of consecutive periods of 10 days with precipitation equal to or less than 1 mm day^{-1} from September to March. Analyzes of frequency were performed by a moving 10 days scales (1-10 / 09, 2-11 / 09, 3-12 / 09, and so on). This procedure avoids the omission of consecutive 10 days scales periods with no rainfall that could occur when considering only the 1-10, 11-20, and 20-30 deciles of each month.

b) Periods ≥ 20 days without rainfall: The occurrence of consecutive periods without rainfall of 20 days or more and the highest droughts periods of each year were verified in the data series. Rainfall days were those in which precipitation was equal to or greater than 1.0 mm. A descriptive analysis of the mean of the events that had 20 days or more without rain, total of days over 20 days without rainfall, and error and standard deviation were also made in these same periods. We calculated the frequencies of the dry periods and the adjustment of the largest annual droughts periods to the distribution of extreme values whose probability density function f (X) and the cumulative probability function F (X) have the following format (ASSIS et al. 1996; COSTA et al., 2009):

$$f(X) = \frac{1}{\beta}\, e^{-\frac{x-a}{\beta}}\, e^{-e^{-\frac{x-a}{\beta}}}$$

$$F(X) = \frac{1}{\beta}\, -e^{-\frac{x-a}{\beta}}$$

where X is the random variable, α is the parameter that controls the position of the curve on the axis of the abscissa and β is the parameter that controls the dimensions of the curve, given a constant shape.

In this study was used Lieblein method to estimate the α and β parameters and the Kolmogorov-Smirnov adhesion test, as described by Assis (1996) and Costa et al. (2009).

The meteorological stations of municipalities Matelândia, Santa Lucia, Terra Roxa and Bom Princípio (Toledo) were used as models.

RESULTS AND DISCUSSION

Rainfall distribution

The mean annual precipitation in the Western of Paraná State (Figure 4) shows a large regional discrepancy. In the extreme Northern and Western, the minimum values are 1600 mm, while in a short distance the East, Central and South portions have maximums of 1950 mm. The rainfall shows a great similarity with the altitude of the region (Figure 02), with the lowest portion of the region from 100 to 300 m in average, near the Paraná River and low Iguaçu, with lower precipitation heights, while the highest portion in the basin of 450 m to 800 m presenting the highest rainfall heights.

Rainfall in the West of Paraná State is affectd by different factos, including continentally, relief forms, altitude and the displacement of systems. The cold fronts are characterized by the encounter of the polar mass air with the mass of continental hot and humid air. This can generate strong atmospheric instabilities, often with formation of cumulonimbus and severe storms, accompanied by strong gusts of wind and hail precipitation, or rain of weak to moderate intensity, but with a duration of days in which the front remains stationary. These phenomena mainly take place in the fall, Winter and Spring (CALDANA et al., 2018; NITSCHE et al., 2019; CALDANA et al., 2019).

Figure 3. Annual rainfall average in Western of Paraná State.

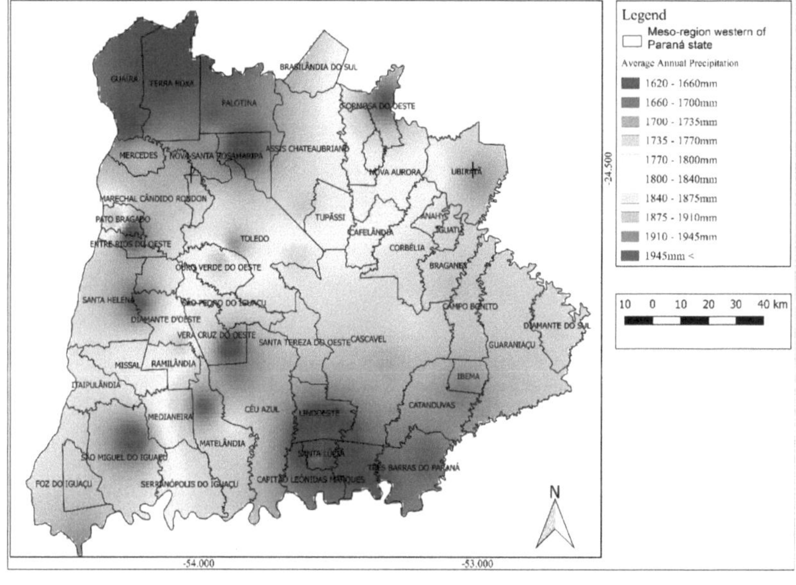

Source: ANA; ÁGUAS PARANÁ; INMET; IAPAR.

The polar mass air has trajectory is favored by the Paraná River and Iguaçu River, having great impact in the rainfall in this region. They are identified satellite images as a large line of instability that advances through Paraná State in the Southeast - Northeast direction. And as observed, the relief in the Southern portion of the region has a rapid rise, from 200 to 900 m (Figure 1); this shock with the relief can bring more rainfall in the higher portion (BEREZUK e SANT'ANNA NETO, 2006; KUNZ et al., 2009; BEREZUK, 2017; PUNGE et al., 2017; SANCHEZ et al., 2017).

This aspect contributes to the greater rainfall precipitations in the areas that the relief ascends in the sense of municipality of Cascavel, both in the direction of the Paraná River (West-East), and in the direction of the Iguaçu River (South-North). In both directions there is a great rise in a short distance, from 200 to 900 m and 400 to 900 m, respectively.

Another two main types of precipitation formation in the region are the Convective Systems and the Mesoscale Convective Complexes (CCM), which operate throughout the year but predominate in the spring and summer seasons. CCMs are identified in satellite images by their approximately circular shape and by a wide area of storm coverage. They are defined as a cluster of cumulonimbus covered by a dense layer of cirrus, and convective cloud systems with rapid vertical and horizontal growth over a period of 6 to 12 hours.

Depending on their intensity, they can create multiple nuclei with storm formation and hail incidence. Its displacement is normally in the direction West - East, coming from Paraguay. The shape of the relief, mountainous areas and altimetric rise over short distances contribute to the elevation of warm and humid air, strengthening this system and causing more rainfall in these areas. The convective systems differ from these by the smaller spatial extent, forming several precipitation nuclei dispersed by the Paraná State (HOLLEMAN, 2001; KUNZ et al., 2009; PUNGE; KUNZ, 2016; DAFIS et al., 2017; PUNGE et al., 2017; TREFAULT et al., 2018; CALDANA et al., 2018; CALDANA et al., 2019).

In this context, the relief also contributes to the formation and intensity of the Convective System, influencing the unequal regional distribution. The location of the state in a climatic transition area contributes to the discrepancies of temperature and, consequently, of pressure, influencing the displacement of the systems (CARAMORI et al., 2001; CALDANA et al., 2018). It is still worth noting that the region is located on the border with Paraguay and Argentina and in the center of South America, not suffering from maritime influence.

To complement the analysis by means of the box plot (Figure 5) it was possible to verify a great discrepancy between the rainfall heights at the Matelândia station compared to the others. The interval between Q1 and Q3 was 1,290 mm to 2,195 mm. The median was 1,975 mm and the maximum value was 3,006 mm, the highest annual rainfall recorded in the series of the analyzed stations.

Even with a maximum rainfall lower than the Matelândia station, the medians of Toledo and Santa Lucia stations showed values close to that of

Matelândia, with 1,860 and 1,805 mm, respectively. The maximum values recorded were also close, with 2,620 and 2,610 mm, respectively.

Figure 4. Annual rainfall variability in Western of Paraná State.

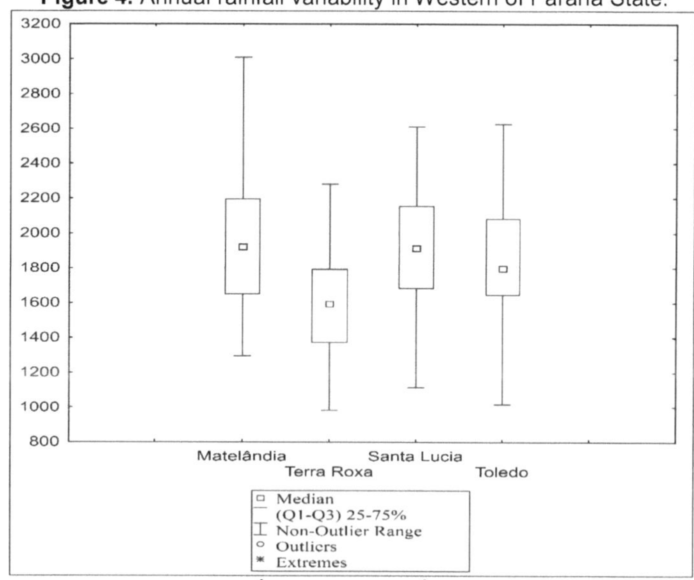

Source: ANA; ÁGUAS PARANÁ; INMET; IAPAR.

The Terra Roxa station presented values with relative dispersion for the others. The median was 1,602 mm and the maximum and minimum values ranged from 2,260 to 960 mm, respectively. The latter was the lowest annual rainfall recorded among the analyzed stations. The interval between Q1 and Q3 of Terra Roxa was 1,798 to 1,385 mm, respectively. Even with the discrepancy between localities, the whole region has been rainy by the annual average, not being a limiting factor for production. Its variation, rhythm and distribution can be limiting aspects.

The monthly variability of the rainfall precipitations in the soybean cycle (Figure 06) presented similar distribution in the region. The Matelândia station featured six discrepant values in the series, one at the bottom of the box plot. In December an extreme value of 505 mm was identified. The medians of these months ranged from 195 mm (October) to 160 mm (September).

The Santa Lucia station presented five discrepant values in the upper part of the box plot, three in the month of January and one in November and December, and a lower value in January. It was observed two precipitation heights below 20 mm monthly in February and December, being the months that presented the greatest variability, with oscillation of 16 to 470 mm.

Figure 5. Rainfall variability during the cycle of soybean crops in Western of Paraná State.

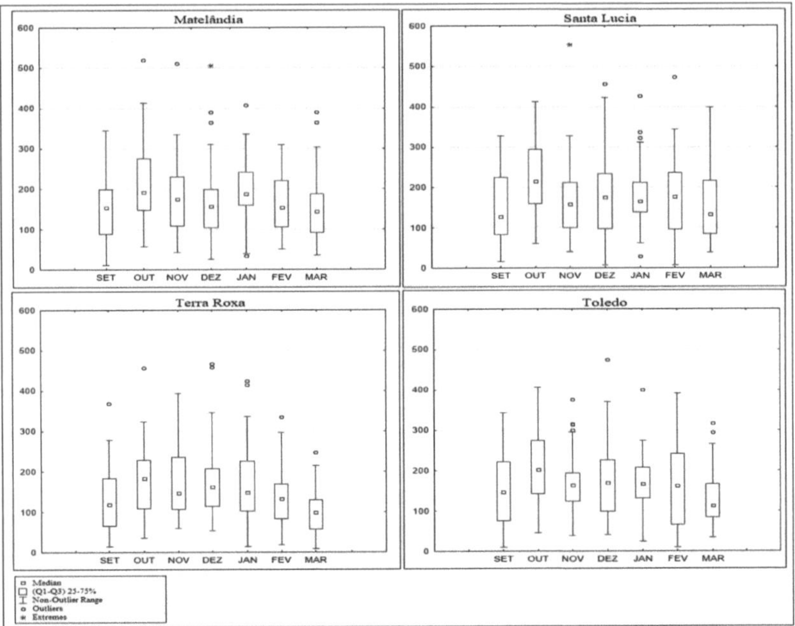

Source: ANA; ÁGUAS PARANÁ; INMET; IAPAR.

For the station of Terra Roxa five discrepant values were identified, all in the part superior of the box plot, being one in the month of October, one in February and two in December and one in January. The station had the lowest precipitation values, with medians of all months below 180 mm.

In the station of Toledo five discrepant values were observed, all in the part superior of the box plot, in the months of November, December and January. Only the month of October had a median of over 200 mm, while in the lower part, the month of February presented the lowest value of monthly precipitation, with 13 mm.

The mean monthly distribution of precipitation was satisfactory for soybean cultivation in the region. The medians remain in all scenarios above 100 mm, and in some months, over 200 mm, but there is a risk of rainfall with discrepant values such as in the lower part of the box plot in Matelândia and Santa Lucia, or, monthly rainfall less than 20 mm, as observed in all seasons, mainly in the month of September.

The 10 days scales of precipitation (Figure 7) showed great variation and numerous discrepant or extreme values. Only five ten-day periods did not register absence of precipitation - Matelândia 01/02; Santa Lucia Decembers 01/11 and 03/12; Land Rover Decay 01/10; Toledo 10 days scales 03/12. The accumulation of consecutive days without rainfall can be determinant for the growth, development and yields of the soybean crop.

The effect of water deficiency on the production of soybeans depends on the intensity and phenological stage in which it occurs (DESCLAUX et al., 2000; BOARD; KAHLON, 2011). The stomatal closure, caused by the dehydration of the cytoplasm, is responsible for the abortion and accelerating the senescence of vegetative and reproductive structures of soybean (CASAGRANDE et al., 2001; BOOTE et al., 2010).

In the reproductive phenologycal stages, if water deficiency coincides with flowering and the beginning of pod formation, abortion of flowers and pods occurs. The final number of pods depends on the vigor of the plants during the flowering period and the weather conditions. However, it is at the stage between pod formation and grain filling (R3-R5) that water deficiency becomes critical. Thus, ten days without rain, depending on the period of development of the crop, can be determinant for the success of the crop (FARIAS et al., 1997; FARIAS et al., 2007; BOARD; KAHLON, 2011; YEE-SCHAN et al., 2013).

Figure 6. Decendial rainfall variability of the soybean cycle months in the Western Region of Paraná State

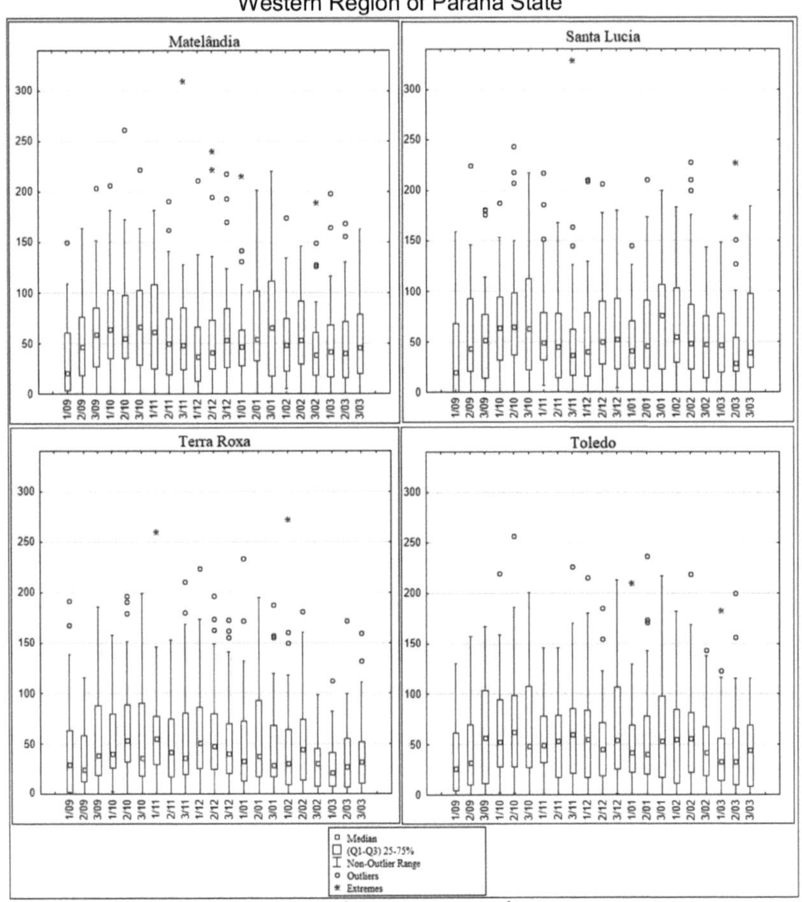

Source: ANA; ÁGUAS PARANÁ; INMET; IAPAR.

The 10 days scales analysis presented great rainfall variability, as for example in the December 3/11, varying from 0 to 312 mm. With the exception of Terra Roxa, the December 01/09 presented the lowest median in the region. As of October, even with great variability, the 10/10 ,10 days scales presented medians greater than 50 mm, denoting greater rainfall regularity.

Analysis of dry periods frequency

10 consecutive days without rainfall

The relative frequencies of moving 10 days dry periods (Figure 8) show variation from 3% to approximately 35% between the months of September and March in the West of Paraná State. The periods with the lowest risk are in the month of October, followed by December 10 to January 5. The biggest risks of dry spells are concentrated in the first half of September, with peaks above 30 %.

Figure 7. Occurrences frequency of dry periods, in decendial between September to March in Western of Paraná State, Brazil.

Source: ANA; ÁGUAS PARANÁ; INMET; IAPAR.

In the period from December to January soybean is in the reproductive phenological stages in this region. In the phases of flowering, formation and filling of grains the soybean has high water requirement. During these two months, the risk varies from 5 to 25%, however, is lower than in the months of November,

February and March. In the month of January there were few occurrences of 10-day dry periods, highlighting the Terra Roxa station with greater risk, while Santa Lucia and Matelândia, located in the South of the region, presented the lowest frequencies.

In the second half of January there is a risk of dry spells ranging from 10 to 20%, but at this time most of the crops in the region are already in physiological maturation, depending on the sowing time, which is carried out between the end September and mid-October, depending on water availability. The risk increases in the months of February and March; however, this is coincident to senescence and harvest.

Periods ≥ 20 days without rain during the year

Most years have at least a period of 20 consecutive days without rain (Figure 10). The season with more occurrences annually was Terra Roxa with 106 events between 1986 and 2017. Only the year 2000 did not have at least 20 days without rain in this locality. It was not possible to identify trends of increase or decrease of these occurrences in the relatively short period of data analyzed.

The station with the lowest number of occurrences was Matelândia, with 53 occurrences during the analyzed series. Four years in the series were identified without any rainfall within 20 days.

Figure 8. Occurrences at least consecutives 20 days with no rainfall in Western of Paraná State, Brazil.

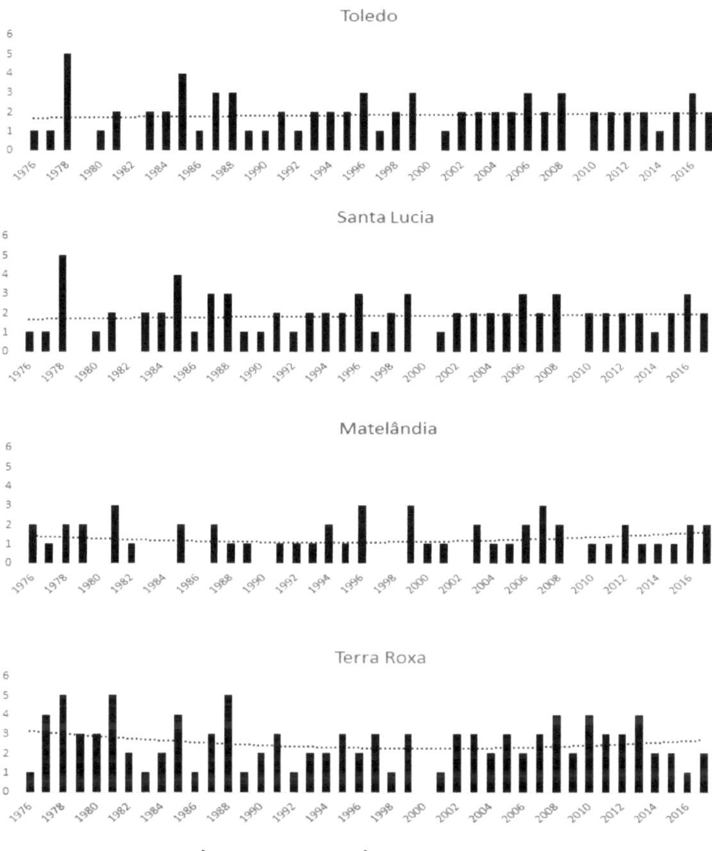

Source: ANA; ÁGUAS PARANÁ; INMET; IAPAR. Adapted and organized by the authors (2019).

The distribution of consecutives days under extremes droughts (Figure 11), was identified that, even in the region with registered high rainfall heights. The risk of reaching at least 20 days without any rainfall is about 80%, while for the other stations is 90 %.

Figure 9. Cumulative probability of the largest annual dry period in the Western Meso-region, according to the distribution of extremes

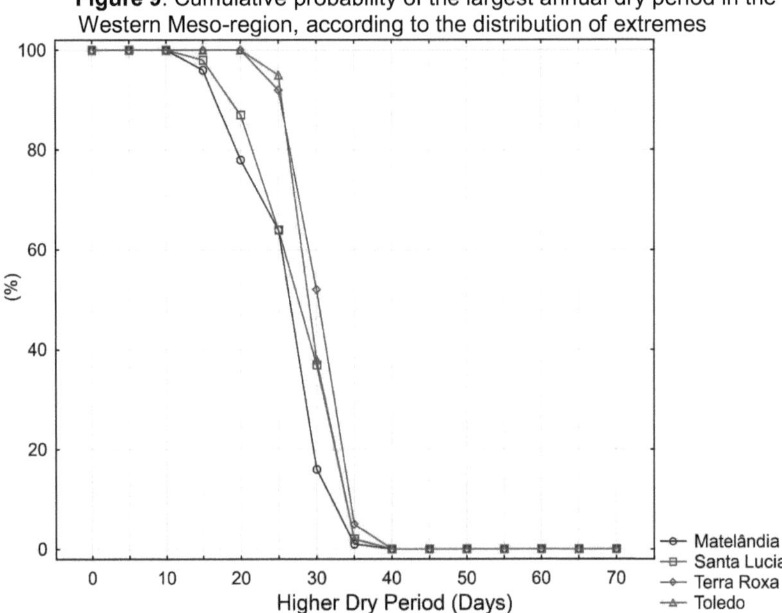

Source: ANA; ÁGUAS PARANÁ; INMET; IAPAR. Adapted and organized by the authors (2019).

The risk tends to zero for durations of 30 to 40 days without rainfall. These occurrences were few in the period that the soybean crops are in the field (Table 01). Three of the occurrences were in March, a late harvest period, presenting no risk for soybeans. One of them, being the most severe one, occurred practically throughout the month of September, with accumulated precipitation of 1.1 mm, indicating once again the importance of adjusting the sowing period for the month of October.

Table 01: Date of occurrence and accumulated rainfall of periods of more than 20 days without rain at the Matelândia station from September to March.

Data	Year	Total Rainfall (mm)
January 09 to 28	1978	0mm
January 10 to 31 de	1982	0mm
February 23 to March 17	1987	0mm
March 06 to 28	1988	8.6mm
March 08 to 31	2002	0mm
November 13 to December 02	2008	0mm
November 23 to December 30	2011	4.7mm
September 01 to 29	2017	1.1mm

Source: Made by the Authors (2019).

This work demonstrates the importance of the agricultural calendar for planning of the soybean crop in the Western of Paraná State, aiming to reduce risks and obtain the highest yields. There is a need to adapt the sowing season, always aiming at the most regular rainy period, corresponding to the month of October. Therefore, the occurrence of the phases of grain filling and physiological maturation in the months of December and January is fundamental to minimize the risks of dry periods that could compromise the physiological and productive yields of soybean.

FINAL CONSIDERATIONS

The average rainfall precipitations in the Western of Paraná State is enough for the soybean crop, however, the rainfall variability is an important cause of production loss. The rainfall distribution is not regular both in the annual scale and in shorter periods during the crop cycle.

During the soybean crop cycle, the highest risk occurs in the months of September, February and March, however in February and March the crop is already in senescence. According to the water requirement of soybean crops, the sowing is recommended in the month of October, to guarantee efficiency in the emergency and establishment. Besides, the period of flowering and grain filling occurs between December and January, where there is also a lower risk of water deficiency in the Western of Paraná State.

The stations located in the lower portion of the region (west, north, and southwest) near the Paraná and Iguaçu rivers presented lower rainfall heights than those located in higher altitudes of the central, east, northeast and south portions. The factor that justifies this discrepancy is the rapid ascent of the relief, both from the Paraná River (West-East) and the lower Iguaçu (South-North), 200-900 m and 400-900 m, respectively. In the municipality of Terra Roxa, located in the portion with lower rainfall of the region, the risks of consecutive dry periods are higher than in the other locations, however, even in the driest portion of the region the conditions of cropping are suitable if the farmer elaborates a climate plan.

REFERENCES

ADEBOYE, O. B. et al. Performance evaluation of AquaCrop in simulating soil water storage, yield, and water productivity of rainfed soybeans (Glycine max L. merr) in Ile-Ife, Nigeria. **Agricultural Water Management**, v. 213, p. 1130-1146, 2019.

AGOVINO, M. et al. Agriculture, climate change and sustainability: The case of EU-28. **Ecological Indicators**, DOI: 10.1016/j.ecolind.2018.04.064. 2018.

AHMAD, M. et al. Modelling soil water dynamics and crop water use in a soybean-wheat rotation under chisel tillage in a sandy clay loam soil. **Geoderma**, v. 327, p. 13-24, 2018.

AKHTAR, K. et al. Wheat straw mulching offset soil moisture deficient for improving physiological and growth performance of summer sown soybean. **Agricultural water management**, v. 211, p. 16-25, 2019.

ALBUQUERQUE, F. da S. et al. Aptidão climática de culturas agrícolas importantes para comunidades indígenas do semiárido brasileiro. **Irriga**, v. 22, n. 1, p. 59, 2018.

ALVARES, C. A. et al. Köppen's climate classification map for Brazil. **Meteorologische Zeitschrift**, v. 22, n. 6, p. 711-728, 2013.

ASSIS, F. N. et al. **Aplicações de estatística à climatologia: teoria e prática.** Pelotas: UFPel, 1996. 161 p.

BATTISTI, R. et al. Assessment of soybean yield with altered water-related genetic improvement traits under climate change in Southern Brazil. **European journal of agronomy**, v. 83, p. 1-14, 2017.

BENCKE-MALATO, M. et al. Short-term responses of soybean roots to individual and combinatorial effects of elevated [CO_2] and water deficit. **Plant Science**, v. 280, p. 283-296, 2019.

BEREZUK, A. G.; SANT'ANNA NETO, J. M. Eventos climáticos extremos no oeste paulista e norte do Paraná, nos anos de 1997, 1998 e 2001. **Revista Brasileira de Climatologia**, v. 2, p. 9-22, 2006.

BEREZUK, A. G. Eventos Extremos: Estudo da Chuva de Granizo de 21 de Abril de 2008 na Cidade de Maringá-PR. **Revista Brasileira de Climatologia**, v. 5, p. 153-164, 2017.

BERLATO, M. A. et al. Evapotranspiração máxima da soja e relações com a evapotranspiração calculada pela equação de Penman, evaporação do tanque "classe A" e radiação solar global. **Agronomia Sulriograndense**, v. 22, n. 2, p. 243-259, 1986.

BERLATO, M. A. et al. Associação entre El Niño Oscilação Sul e a produtividade do milho no Estado do Rio Grande do Sul. **Pesquisa Agropecuária Brasileira**, v. 40, n. 5, p. 423-432, 2005.

BOARD, J. E.; KAHLON, C. S. Soybean yield formation: what controls it and how it can be improved? In: EL-SHEMY, H. A. **Soybean physiology and biochemistry.** Croatia: In Tech, p. 1-36. 2011.

BOOTE, K. J. et al. Testing effects of climate change in crop models. In: HILLEL, D.; ROZNZWEIG, C. **Handbook of climate change and agroecosystems**. London: Imperial College Press, 2010. p. 109-129.

CALDANA, N. F. S. et al. Ocorrências de Alagamentos, Enxurradas e Inundações e a Variabilidade Pluviométrica na Bacia Hidrográfica do Rio Iguaçu. **Revista Brasileira de Climatologia**, v. 23, p. 343-355, 2018.

CALDANA, N. F. S. et al. Gênese, Impacto e a Variabilidade das Precipitações de Granizo na Mesorregião Centro-Sul Paranaense, Brasil. **Caderno De Geografia**, v. 29, p. 61-80, 2019.

CARAMORI, P. H. et al. Zoneamento de riscos climáticos para a cultura do café (*Coffea arabica* L.) no Paraná. **Revista Brasileira de Agrometeorologia**, Santa Maria, v. 9, n. 3, p.486-494, 2001.

CARAMORI, P. H. et al. Sistema de alerta para geadas na cafeicultura do Paraná. **Informe Agropecuário (Belo Horizonte)**, v. 28, p. 66-71, 2007.

CARAMORI, P. H. et al. Agrometeorologia operacional no estado do Paraná. **Agrometeoros**, Passo Fundo, v.24, n.1, p.65-70, 2016.

CASAGRANDE, E. C. et al. Expressão gênica durante déficit hídrico em soja. **Revista Brasileira de Fisiologia Vegetal**, v. 13, n. 2, p. 168-184, 2001.

CONCEIÇÃO, M. A. F. Índices hídricos climáticos para a viticultura. **Pesquisa Agropecuária Brasileira**, v. 53, n. 6, p. 765-768, 2018.

COSTA, A. B. F. et al. **Análise climatológica de dias consecutivos sem chuva no Estado do Paraná.** In: III Simpósio Internacional de Climatologia, 2009, Canela - RS. Mudanças de Clima e Extremos e Avaliação de riscos futuros, planejamentos e desenvolvimento sustentável, 2009.

DAFIS, S. et al. Observational and modeling study of a mesoscale convective system during the HyMeX—SOP1. **Atmospheric Research**, v. 187, p. 1-15, 2017.

DERAL. Departamento de Economia Rural do Estado do Paraná. **Tabela de Produção Agrícola por Município**. Disponível em < http://www.agricultura.pr.gov.br/modules/conteudo/conteudo.php?conteudo=137 > Acesso em 14/02/2019.

DESCLAUX, D. et al. Identification of soybean plant characteristics that indicate the timing of drought stress. **Crop science**, v. 40, n. 3, p. 716-722, 2000.

DIETZEL, R. et al. How efficiently do corn-and soybean-based cropping systems use water? A systems modeling analysis. **Global change biology**, v. 22, n. 2, p. 666-681, 2016.

DU, J. et al. Maize-soybean strip intercropping: Achieved a balance between high productivity and sustainability. **Journal of integrative agriculture**, v. 17, n. 4, p. 747-754, 2018.

ELY D. F.; DUBREUIL, V. Análise das Tendências Espaço-Temporais das Precipitações anuais para o Estado do Paraná - Brasil. **Revista Brasileira de Climatologia**, v. 21, n. 13 p. 553-569, 2017.

FARIAS, J. R. B., et al. Zoneamento agroclimático da cultura da soja para o estado do Paraná. **EMBRAPA-CNPSo**, 1997.

FARIAS, J. R. B. et al. Ecofisiologia da soja. **Embrapa Soja-Circular Técnica (INFOTECA-E)**, 2007.

GAJIĆ, B. et al. Effect of irrigation regime on yield, harvest index and water productivity of soybean grown under different precipitation conditions in a temperate environment. **Agricultural Water Management**, v. 210, p. 224-231, 2018.

GARCIA, R. A. et al. Soybean-corn succession according to seeding date. **Pesquisa Agropecuária Brasileira**, v. 53, n. 1, p. 22-29, 2018.

GELCER, E. et al. Influence of El Niño-Southern oscillation (ENSO) on agroclimatic zoning for tomato in Mozambique. **Agricultural and Forest Meteorology**, v. 248, p. 316-328, 2018.

GIMÉNEZ, L.; et al. Water use and yield of soybean under various irrigation regimes and severe water stress. **Application of AquaCrop and SIMDualKc models. Water**, v. 9, n. 6, p. 393, 2017.

GRASSINI, P. et al. Soybean yield gaps and water productivity in the western US **Corn Belt. Field Crops Research**, v. 179, p. 150-163, 2015.

HOLLEMAN, I. **Hail Detection Using Single-Polarization Radar.** Ministerie van Verkeer en Waterstaat, Koninklijk Nederlands Meteorologisch Instituut, 2001.

JACONDINO, W. D. et al. Análise de Veranicos Intensos na Região Sul do Brasil e Condições Sinóticas Associadas. **Anuário do Instituto de Geociências**, v. 41, n. 2, 2018.

JHA, P. K.; et al. Responses of soybean to water stress and supplemental irrigation in upper Indo-Gangetic plain: Field experiment and modeling approach. **Field crops research**, v. 219, p. 76-86, 2018.

JÚNIOR, R. de S. N.; SENTELHAS, P. C. Soybean-maize succession in Brazil: Impacts of sowing dates on climate variability, yields and economic profitability. **European Journal of Agronomy**, v. 103, p. 140-151, 2019.

KARIMI, V. et al. Climate change and agriculture: Impacts and adaptive responses in Iran. **Journal of Integrative Agriculture**, v. 17, n. 1, p. 1-15, 2018.

KHOJELY, D. M. et al. History, current status, and prospects of soybean production and research in sub-Saharan Africa. **The Crop Journal**, v. 6, n. 3 p. 226-235, 2018.

KUNZ, M. et al. Recent trends of thunderstorm and hailstorm frequency and their relation to atmospheric characteristics in southwest Germany. **International Journal of Climatology: A Journal of the Royal Meteorological Society**, v. 29, n. 15, p. 2283-2297, 2009.

LEM, S. et al. The heuristic interpretation of box plots. **Learning and Instruction**, v. 26, p. 22-35, 2013.

LUNGARSKA, A.; CHAKIR, R. Climate induced land use change in France: impacts of agricultural adaptation and climate change mitigation. **Ecological Economics**, v. 147, p. 134-154, 2018.

DE MATOS, R. M. et al. Agroclimatic Potential For Mango Culture in the Municipality of Barbalha CE. **Revista Brasileira de Agricultura Irrigada**, v. 12, n. 1, p. 2366, 2018.

MEDEIROS, G. B.; ZANÃO JUNIOR, L. A. Agricultura e Produtividade da Soja no Entorno do Reservatório de Itaipu. In: ZANÃO JUNIOR, et al. **Produtividade da Soja no Entorno do Reservatório de Itaipu**. IAPAR. 2015.

MEDEIROS, R. M. et al. Aptidões climáticas: caju, palma forrageira e milho no município de São Bento do Una-PE, Brasil. **Journal of Environmental Analysis and Progress**, v. 3, n. 3, p. 310-318, 2018.

MINUZZI, R. B.; CARAMORI, P. H. Variabilidade climática sazonal e anual da chuva e veranicos no Estado do Paraná. **Ceres**, v. 58, n. 5, 2015.

MUELLER, T. G. et al. Map quality for ordinary kriging and inverse distance weighted interpolation. **Soil Science Society of America Journal,** v. 68, n. 6, p. 2042-2047, 2004.

NITSCHE, P. R. et al. **Atlas Climático do Estado do Paraná.** Instituto Agronômico do Paraná. Disponível em: <http://www.iapar.br/>. Acesso em: 23 de fevereiro de 2019.

PATHMESWARAN, C. et al. Impact of extreme weather events on coconut productivity in three climatic zones of Sri Lanka. **European Journal of Agronomy**, v. 96, p. 47-53, 2018.

PEDERSEN, P.; LAUER, J. G. Response of soybean yield components to management system and planting date. **Agronomy Journal**, v. 96, n. 5, p. 1372-1381, 2004.

PUNGE, H. J.; KUNZ, M. Hail observations and hailstorm characteristics in Europe: A review. **Atmospheric Research**, v. 176, p. 159-184, 2016.

PUNGE H. J. et al. Hail frequency estimation across Europe based on a combination of overshooting top detections and the ERA-INTERIM reanalysis. **Atmospheric Research**, v. 198, p. 34-43, 2017.

DOS REIS, L. S. et al. Zoneamento agroclimático de cultivos com potencial energético no estado de Minas Gerais. **Revista Espinhaço**, p. 50-63, 2017.

SANCHEZ, J. L. et al. Are meteorological conditions favoring hail precipitation change in Southern Europe? Analysis of the period 1948–2015. **Atmospheric Research,** v. 198, p. 1-10, 2017.

SANT'ANNA NETO, J. L. Weather Lore de Pindorama: o conhecimento sobre o tempo e o clima no período não instrumental na antiguidade e no Brasil pré-cabralino. **Iberografias**, v. 11, p. 112-121, 2015.

SANTI, A. et al. Impacto de cenários futuros de clima no zoneamento agroclimático do trigo na região Sul do Brasil. **Agrometeoros**, v. 25, n. 2, p. 303-311, 2018.

SENTELHAS, P. C. et al. The soybean yield gap in Brazil–magnitude, causes and possible solutions for sustainable production. **The journal of agricultural science**, v. 153, n. 8, p. 1394-1411, 2015.

SILVA, A. C. da; et al. A importância da soja para o agronegócio brasileiro: uma análise sob o enfoque da produção, emprego e exportação. **V Encontro de Economia Catarinense**, 2011.

SILVESTRE, M. R. et al. Critérios estatísticos para definir anos padrão: uma contribuição à climatologia geográfica. **Revista Formação**, v. 2, n. 20, p. 23-53, 2013.

SCHNEIDER, H.; DA SILVA, C. A. O uso do modelo box plot na identificação de anos-padrão secos, chuvosos e habituais na microrregião de Dourados, Mato Grosso do Sul. **Revista do Departamento de Geografia**, v. 27, p. 131-146, 2014.

DE SOUZA, D. C. F. et al. Zoneamento Agroclimático da Palma Forrageira (Opuntia Sp) Para o Estado de Sergipe. **Revista Brasileira de Agricultura Irrigada**, v. 12, n. 1, p. 2338, 2018.

DE SOUSA, J. W. ; DE OLIVEIRA, P. F. Risco climático para o café Conilon (Coffea canephora) nos municípios de Rio Branco, Tarauacá e Cruzeiro do Sul, AC. **Revista Brasileira de Ciências da Amazônia/Brazilian Journal of Science of the Amazon**, v. 7, n. 2, p. 31-40, 2018.

STÜLP, M. et al. Agronomic traits and seed yield produced in the soybean-corn crop in succession cropping. **Acta Scientiarum. Agronomy**, v. 32, n. 4, p. 651-661, 2010.

TAYT'SOHN, F. C. O. Assessing sugarcane expansion to ethanol production under climate change scenarios in Paranaíba river basin–Brazil. **Biomass and Bioenergy**, v. 119, p. 436-445, 2018.

TREFALT, S. et al. A Severe Hail Storm in Complex Topography in Switzerland-Observations And Processes. **Atmospheric Research**, v. 209, p. 76-94, 2018.

WIRÉHN, L. Nordic agriculture under climate change: A systematic review of challenges, opportunities and adaptation strategies for crop production. **Land use policy**, v. 77, p. 63-74, 2018.

YEE-SHAN, K, et al. Drought stress and tolerance in soybean. In: BOARD, J.E. A comprehensive survey of international soybean research:

genetics, physiology, agronomy and Nitrogen Relationship. **Croatia: In Tech**, p. 208-237, 2013.

ZARO, G. C. et al. Inter-annual analysis of soybean and corn yield in relation to water deficit in a transitional zone between subtropical and tropical climate. **Australian Journal of Crop Science**, v. 12, n. 4, p. 511, 2018.

Printed by Books on Demand GmbH, Norderstedt / Germany